PEOPLE, SHEEP AND NATURE CONSERVATION

PEOPLE, SHEEP AND NATURE CONSERVATION

THE TASMANIAN EXPERIENCE

Editors: Jamie Kirkpatrick and Kerry Bridle

CSIRO
PUBLISHING

National Library of Australia Cataloguing-in-Publication entry

People, sheep and nature conservation: the Tasmanian experience.

Bibliography.
Includes index.
ISBN 9780643093720.

1. Environmental protection – Tasmania.
2. Sheep industry – Environmental aspects – Tasmania.
3. Sheep industry – Tasmania.
I. Kirkpatrick, J. B. (James Barrie).
II. Bridle, Kerry Lynn.
III. CSIRO.

333.731609946

Published by
CSIRO PUBLISHING
150 Oxford Street (PO Box 1139)
Collingwood VIC 3066
Australia

Telephone: +61 3 9662 7666
Local call: 1300 788 000 (Australia only)
Fax: +61 3 9662 7555
Email: publishing.sales@csiro.au
Web site: www.publish.csiro.com

Front cover photo by Diana Cameron
Back cover photo by Jamie Kirkpatrick

Set in 11/14 Adobe Minion and ITC Stone Sans
Edited by Anne Findlay
Cover and text design by James Kelly
Typeset by Desktop Concepts Pty Ltd, Melbourne
Printed in Australia by BPA Print Group

Contents

Preface

J.B. Kirkpatrick and K.L. Bridle

Runs are areas of natural and semi-natural vegetation where their owners or leaseholders run stock. Much of the wool produced in the world comes from runs, whether in Patagonia, Peru or the Midlands of Tasmania.

The focus of the book is on the interactions between people, sheep and nature in the run country of Tasmania. It addresses the following questions: *What is the relationship between production and conservation in natural landscapes grazed by sheep (*Ovis aries*)? Can nature conservation be integrated with wool production?* The book relates to the wool-growing and nature conservation industries. It is most accurately described overall as environmental geography, although it incorporates disturbance ecology, environmental history, cultural ecology, political ecology, social geography, rural sociology and economic geography.

Private land is the new frontier for nature conservation in Australia. The question *How can we ensure that the many conservation values that rely on private land will be perpetuated?* has vexed many of Australia's greatest minds, and has been the focus of major expenditure by governments and landowners, as in the Landcare movement and the Natural Heritage Trust. One of the major tribes on the frontier is the wool-growing community, which utilises large areas of conservation-significant natural and semi-natural vegetation, their runs, to produce a product that has gone from the foundation of Australia's wealth to an economically important rural activity with poor or negative returns on capital.

A major purpose of the book was to search for potentially effective pathways to maintain conservation values in a production context, ways that would suit the cultural, social and economic needs of wool-growing communities. This quest required an understanding of the ways that wool growing evolved on the runs, the ways in which graziers utilise and value their runs within their enterprises, the ways in which sheep, under different management regimes, affect natural values, and the ways in which the cultural, social and political values of wool growers and conservationists can be reconciled. The achievement of this understanding fulfils two extra purposes: to communicate the manifold ways in which graziers solve the ecological, social and economic problems of their enterprises; and, to exemplify the complexities and contingencies of the reciprocal interactions between people, their stock and native nature.

In achieving our aims we have used a wide variety of research approaches. We report the results of research in the classic hypothesis-testing mode, in descriptive scientific mode, in the mode of normative social science, and in qualitative mode. These are complementary and synergistic approaches in any quest to understand relationships between people and nature, although they are most often viewed as antagonistic.

Chapter 1 is a history of the relationships between people, sheep and nature in Tasmania, approached more systematically than temporally. Chapters 2 and 3 are largely based on material from interviews with the managers of 48 wool-producing properties, material that is analysed in both normative and qualitative modes. The second chapter describes the ways in which graziers derive income from, and manage, their runs, in the context of their property as a whole. The third chapter describes the ways wool producers feel about their runs and how they conserve these areas. Trees turned out to be a major focus of conservation concern for graziers, so Chapter 4 reports scientific investigations of changes in tree cover, changes in tree species composition and the relative effectiveness of techniques for tree establishment. A scientific approach is also used in Chapter 5, which reviews the ways in which sheep and property management influence nature on the runs. Chapter 6 addresses the major conservation problem of the runs, their ongoing disappearance and the consequences of their fragmentation. In Chapter 7 we look at the history of the interaction of graziers with conservation bureaucrats, the attributes of a fair and effective conservation system for the runs, the ways in which the nature of the runs can be used in marketing wool, and the overall potential for integrating profitable use of the runs for wool with nature conservation. A brief final chapter, Chapter 8, draws out some general themes.

Details of methodology and statistical results have been delegated to notes, in small print, at the end of chapters. Each of the chapters has been reviewed by a Tasmanian wool producer, as well as a relevant expert, to help ensure accuracy, comprehensibility and comprehensiveness.

The chapters in the book involve many co-authors, all of whom undertook the work reported in the book while they were students or staff in the School of Geography and Environmental Studies, University of Tasmania. The new research that is reported in the book has been supported by Environment Australia, the Land and Water Resources Research and Development Corporation, the Australian Research Council (DP0665083), the Department of Primary Industries and Water, the Parks and Wildlife Service, the University of Tasmania, and, most substantially, the Native Vegetation and Biodiversity Sub-program of Land, Water & Wool Program of Australian Wool Innovation and Land and Water Australia. Land, Water & Wool, through a grant to the senior author, supported the writing and publication of this volume, as well as most of the research reported herein.

To ensure continuity in style and argument, the senior author has reworked the contributions of the other authors of the book and chapters. Both authors were responsible for editorial work. We are immensely grateful to the many people, largely wool producers, who have been involved in the various projects reported in this book. The reviewers included Jann Williams, Andrew Johnston, David and Brenda Atkinson, Henry Foster, Simon Foster, Bill Gibson, Ian Rogan, Lou Hogan, Mike Wagg, Biz Nicolson, Andrew

Cameron, Kenneth von Bibra, and Tom and Cynthia Dunbabin. The stakeholder group responsible for developing the biodiversity plan was Tom Dunbabin, Henry Foster, Simon Foster, Julian von Bibra, Julian Cotton, Rod Headlam, Sarah Ackland and Steve Barrington, Paul and Shauna Ellis, Meg and Philip Nichols, Lindsay and Rae Young, Eric Hutchinson, Michael Askey-Doran, Andrew Hamlet and Andrew Johnston. There are many other people who have helped in the process of its creation. These include John Cameron, Leanne Sherriff, Kevin Leeson, David Green, Grant Daniels, Steven Leonard and the wool growers quoted so widely in the book. Nevertheless, the opinions expressed in this book, and any errors of fact or interpretation, are those of the relevant authors, not those of Land, Water & Wool or any other party.

We thank the Tasmanian Museum and Art Gallery, Bruce Chetwynd, Kate Ackland, Bill Gibson, Diana Cameron, Matt Appleby, Andrew Cameron and Simon Cameron for providing images used in the book.

Maps

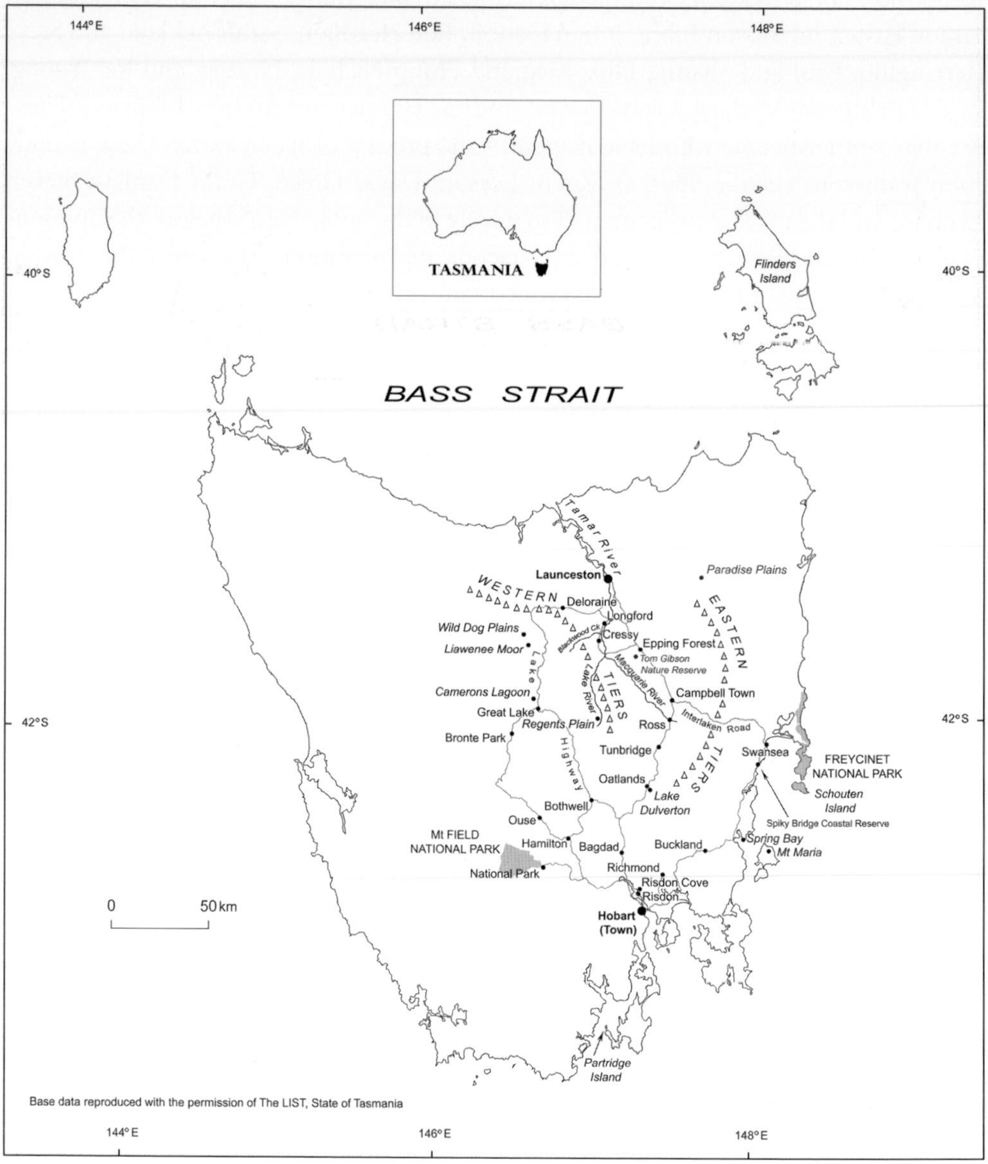

Figure A. Locations mentioned in the text.

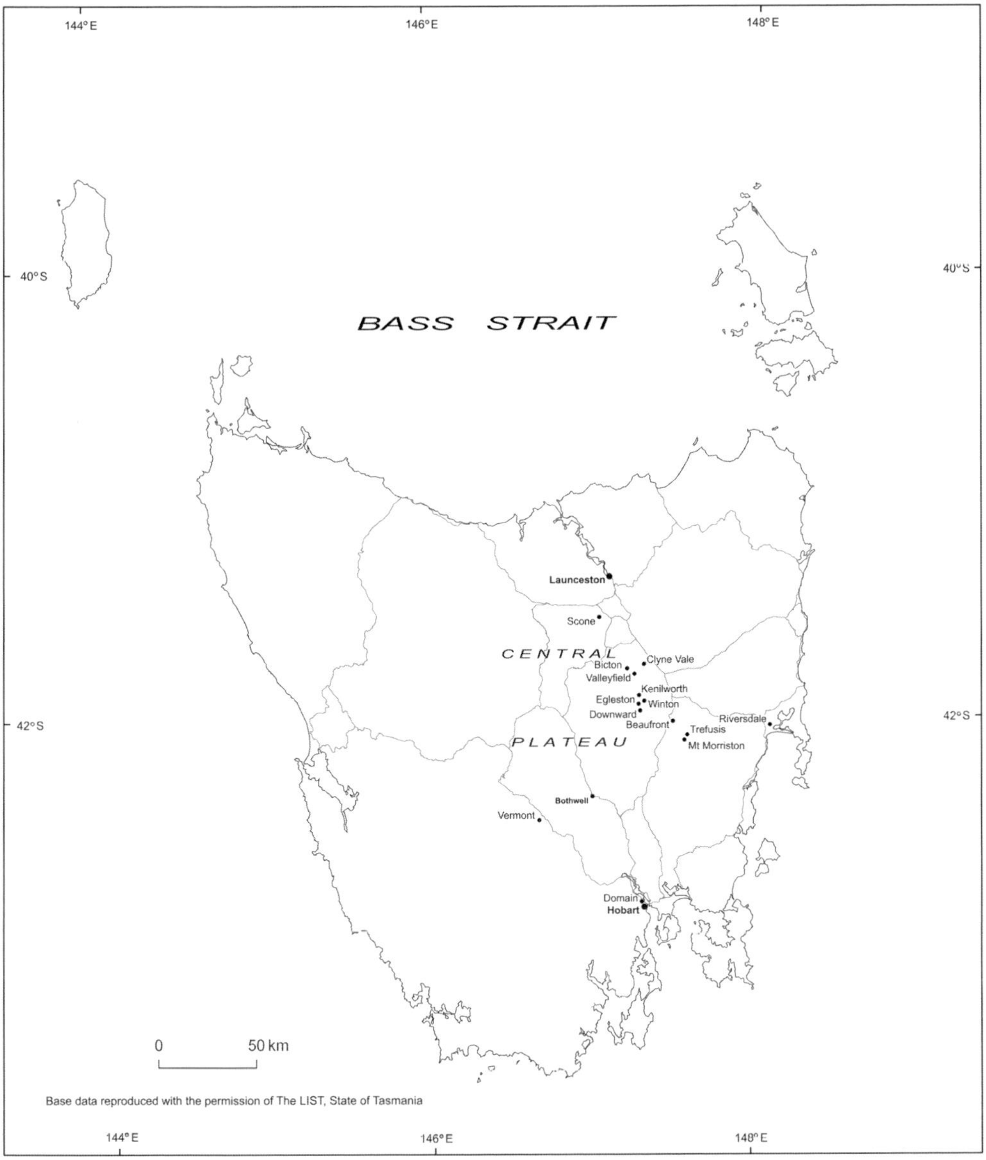

Figure B. Properties mentioned in the text.

Chapter 1

History

J.B. Kirkpatrick

Summary

Sheep first nibbled the native grasses of Tasmania, at Risdon, in 1803. In the mid-nineteenth century a steep growth in their numbers ceased. The growth had been largely unaffected by thylacines, the Black War, poaching and other forms of early colonial larceny. However, with the attainment (or more) of carrying capacity, scab and fluke beset sheep, and rabbits beset the runs. Despite these challenges, graziers bred world-renowned Merinos fit for ecological and economic purposes and devised land management and stock strategies to maintain their pastures. A flirtation with the genes of the wrinkly Vermonts, the 1890s drought, and the introduction and take-off of the green blowfly on the mainland of Australia almost killed the Tasmanian sheep breeding industry. By the early twentieth century the largely native pastures were degraded, scattered with gorse and rippling with rabbits. In the mid-twentieth century a convergence of technologies for land clearance, pasture improvement and pest control, with high prices for wool, led to a linked surge in wool production and the clearance of large areas of native ecosystems. Native animal and plant species responded variously to all these changes as did the incidence of fire in the runs.

Introduction

To understand the present, and make guesses about the future, we must try to understand the past. There has been great change in people, sheep and nature since the first sheep nibbled its first leaf from the native vegetation of Tasmania in 1803. Since this date, people have modified their land management and the genotypes of their sheep to adapt to their local environments, markets and new technologies. Some species of the grasslands and grassy woodlands, like the thylacine (*Thylacinus cynocephalus*) and the

Tasmanian emu (*Dromaeius novaehollandiae diemenensis*) are now extinct, partly because of these changes. Others, like brushtail possums (*Trichosurus vulpecula*), white cockatoos (*Cacatua galerita*) and corbie grubs (*Oncopera* spp.) have benefited from them, to such a degree that they are now regarded as pests. The variability of climate has influenced both economy and ecology, as have intentional and unintentional introductions of alien organisms. This chapter documents the history of the complex interactions between Tasmanian sheep and nature, and the people who made decisions about them.

The grazing lands in gathering and hunting times

In eastern and central Tasmania today, most of the wilder places have little grass or other herbage suited to sheep. Most of the ground cover is usually dominated by aromatic small-leaved shrubs, coarse sedges (Cyperaceae) and saggs (*Lomandra longifolia*). However, there are still some places where a sense can be gained of the type of widespread grassy landscapes, then called 'plains', that elicited enthusiasm from some early European observers. Bowen described the area around Risdon Cove in 1803 as 'more like a nobleman's park than an uncultivated country' (Robson 1983, p. 35). Lt Governor Collins wrote to Viscount Castlereagh in 1806: 'the adjacent country abounds with fine grazing land, that is well clothed with different kinds of grasses' (Jetson 1989, p. 13). Oxley, presumably referring to the Midlands in 1810, in *Remarks on the Country and Settlements formed in Van Diemen's Land 1810* wrote: '… this Rich Interior country will afford Pasturage, and contain at no distant Period Flocks and Herds not inferior in size or Number to the celebrated Plains of Paraguay' (p. 574). Edmund Curr (1824), referring to the Government Stock Farm in the northern Midlands, said:

> 'I can compare this tract to nothing but an Englishman's park. The same variety of wood and lawn is here met with, scattered by the hand of nature, innumerable lagoons are spread in every direction through the whole of this country …' (Historical Committee of the National Trust of Australia (Tasmania) 1966, p. 93).

Tasmania did not have the extensive lowland tussock grassland found on the basalt plains of south-eastern mainland Australia, although the historical evidence does suggest that limited areas occurred in the drier parts of the northern Midlands (Fensham 1989). Many valley bottoms, with dark, cracking clay soils, appear also to have been treeless, except on the edge of streams. The rounded tops of dolerite hills, where the soil was too shallow to support trees, were also covered with grassland (Kirkpatrick 2003). George Augustus Robinson's journals (Plomley 1966) suggest that kangaroo grass (*Themeda triandra*) swards with scattered honeysuckle (*Banksia marginata*) trees covered much of the lime-rich coastal dunes of the far north-west of the state. The larger part of the grassy country had scattered umbrageous trees, as depicted in many early paintings (Figure 1.1), closing into forest on higher ground and on sandier soils (Kirkpatrick 2003).

Many of the trees in the wool-producing landscape of today established well before 1803, in a landscape likely to have been heavily grazed by forester kangaroo (*Macropus giganteus tasmaniensis*), Bennett's wallaby (*Macropus rufogriseus*) and wombat (*Vombatus ursinus tasmaniensis*). Such heavy marsupial grazing in remnant lowland native pastures today produces a dark reddish green sward between the spreading trees, called 'lawn' by Edmund Curr and others. The indigenous people, as far as can be gauged from extremely limited information, were not in great numbers in the lowland grassy country, it being suggested that between the mouth of the Tamar River and the present location of Oatlands there were but 500 people out of an island total of 5000–7000 (Kee 1990). They dug for edible roots and tubers, collected eggs and hunted. They may have burned as frequently as the growth left by marsupials would allow. The soils, dug by people and animals, and uncompressed by hooves, were likely to have been friable and humus-rich.

Over 600 m above sea level kangaroo grass (*Themeda triandra*) gave way to various species of tussock grass (*Poa*), which produced more silvery than red pastures. On basaltic soils and alluvial deposits at higher altitudes, there were extensive grasslands as well as smaller areas of grassy woodland, some with large trees of the much-appreciated alcohol-producing cider gum (*Eucalyptus gunnii*). Some areas of tussock grassland, such as Paradise Plains, may have been created by the burning of rainforest by indigenous people (Ellis and Thomas 1998). Others, such as Liaweenee Moor (Figure 1.2), have been treeless for at least 22 000 years (Thomas and Hope 1994), with trees unlikely to have been excluded by human activity (Fensham and Kirkpatrick 1992). Aboriginal people camped in the forests near these plains (Thomas and Hope 1994). They hunted wallaby that fed on the plains, and, presumably, harvested the many edible grassland plants. There are several historical records of patch burning by indigenous people in these ecosystems (Thomas 1992), and indirect evidence of such burning in the presence of charcoal in swamp and lake sediments (Thomas and Hope 1994).

Thus, in 1803 there were extensive areas of grassland and grassy woodland in Tasmania (Figure 1.3), all well suited to immediate occupancy by European stock, but all also a major source of sustenance for the indigenous people.

The first sheep and their people

The first sheep to be introduced to a land of marsupial-grazed pastures bore little resemblance to most of the sheep found in Tasmania today. In September 1803, 16 ewes, four ewe lambs and three wether lambs were taken ashore at Risdon Cove (Bell 1965, p. 14). The early Tasmanian sheep seem likely to have been crossbred Bengal and Cape breeds. The Cape sheep had fat tails and coarse, often dark, wool (Figure 1.4). The Cape ewe tails tended to get in the way of non-Cape rams (Massy 1990). The Bengals were small sheep with thin hairy coats and thin tails (Historical Committee of

Figure 1.1a. 'Cawood on the Ouse River' by John Glover 1835 (reproduced with the permission of the Tasmanian Museum and Art Gallery).

Figure 1.1b. Cawood on the Ouse River in the early twenty-first century (Bruce Chetwynd).

Figure 1.1c. 'Mills Plains' by John Glover 1836 (reproduced with the permission of the Tasmanian Museum and Art Gallery).

Figure 1.1d. Mills Plains in the early twenty-first century (Bruce Chetwynd).

Figure 1.1e. 'Apsley Park' in 1895 (reproduced with permission from the collection of Kate Ackland).

Figure 1.1f. 'Apsley Park' in the early twenty-first century (Kerry Bridle).

Figure 1.2. Camerons Lagoon on Liaweenee Moor 2006 (top) and Liaweenee Moor grazing exclusion plot and sheep-grazed vegetation 2006 (bottom) (Jamie Kirkpatrick).

the National Trust of Australia 1966, p. 94; Lester 1994). The Bengal had two layers in its coat, the fleece close to the skin, hair above. The Bengal sheep was noted for its prolific breeding, unlike the tail-encumbered Cape sheep. These sheep were destined for the table. Their fleece was regarded almost as a nuisance, being either removed and left to rot, burned or used for filling mattresses (West 1852; Lester 1994).

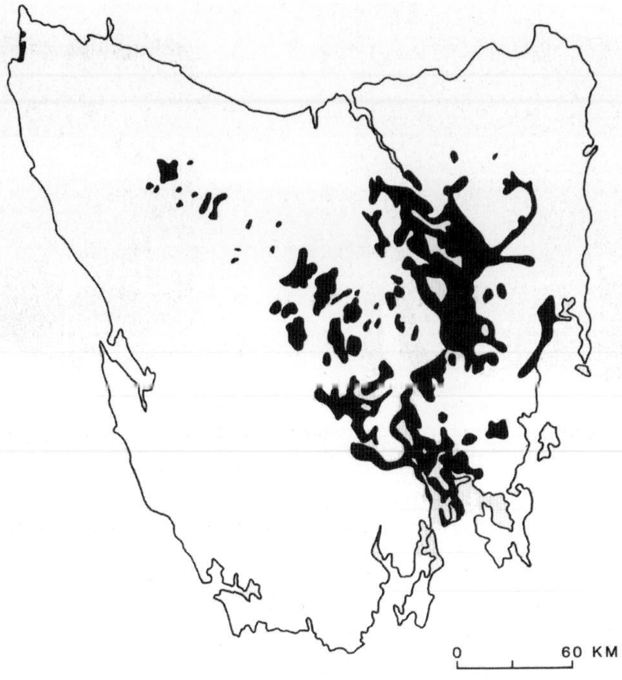

Figure 1.3. Grasslands and grassy woodlands in Tasmania in 1800 (Kirkpatrick *et al.* 1988).

In the first few years of Hobart Town, macropods were more common on the table than mutton (Boyce 2006). By May 1809 a government census of stock in Tasmania reported only 1091 sheep, 122 of which were owned by the government and 121 by Edward Lord. All sheep grazed in Hobart or nearby (Jetson 1989). By 1809 the bloodlines of some of these sheep may have benefited from a few Merino rams 'near to the Spanish type' (Figure 1.4b) sent by Governor King to Hobart in 1805 with the express aim of improving the fleece (Lester 1994). Good breeding stock for wool sheep remained scarce. In 1820 Lieutenant Governor Sorell purchased 300 rams from James Macarthur of Camden, New South Wales, 115 of which died on the voyage or soon thereafter. At this time, James Cox of 'Clarendon', the son of a Merino stud farmer in New South Wales, was a keen buyer of imported rams. His most interesting purchase was a lucky ram imported from Saxony, called, after its importer, Newton. This ram:

> '... was the only animal left alive after bad weather struck the ship causing a very long trip. Fodder ran out and one by one the sheep died. The ram was kept alive on ship's biscuits that so heated its blood that its fleece fell off. The animal arrived in Hobart wrapped in canvas. James Cox, who was in Hobart at the time, heard of this ram, saw the loose fleece and straightaway bought the animal.' (Evandale Historical Society and the Clarendon Committee 1992, p. 32)

Figure 1.4. A Cape sheep (left) and an early nineteenth century Saxon Merino (right).

The first export of Tasmanian wool to London took place in 1822 through the agency of a Mr Henry Hopkins (West 1852). Wool, with its high value, small volume and resistance to decay in ships' holds, was an ideal export at a time when several months might elapse between export from Tasmania and arrival in London. The market at the time was strong, with English looms demanding more raw material than could be gained from the European continent. Between 1827 and 1831 the weight of wool exported to Britain from Van Diemen's Land increased from 192 075 to 1 359 203 pounds (Bischoff 1832).

Sheep and the Black War

By the time of the first ex-Australian export of wool, domestic and feral sheep and cattle were grazing most of the grassy country of eastern and central Tasmania. Kangaroo hunters and shepherds, mostly convicts, were the first Europeans to work in this country (Boyce 2006). Food needed to be obtained for the people in the two major settlements. Officers were given the right to supply kangaroo meat to the government stores. They, in turn, sent out convicts to do the job (Boyce 2006).

In the decade and a half before the process of massive alienation of land commenced, shepherds, and indeed the town dwellers, appeared to have established a *modus vivendi* with the Aboriginal people who took to dogs, tea and sugar very rapidly (Boyce 2006). Many Aborigines learned the English language and wore European clothes (Reynolds 2004). Despite atrocities visited upon them by some sealers, convicts and bushrangers, and early resistance to European poaching of their game (Robson 1983; Boyce 2006), the natives mixed freely with the colonists.

In 1817, only 25 000 acres of land were granted (Parsonson 1998). Most of the Midlands and the central east coast were granted as freehold in the 1820s. In 1821, there were freehold grants of 270 000 acres (Parsonson 1998). In 1823 the peak figure for alienation of Aboriginal/Crown Land was attained, 430 000 acres.

Reynolds (2004), following West (1852), suggested that individual grievances related to atrocities visited upon Aborigines were not enough to cause general resistance to the European invasion, but that this privatisation and partitioning of the land in the early 1820s was the cause of the 'Black War'.

> 'Land unfenced, and flocks and herds moving on hill and dale, left the motions of the native hunters free; but the hedges and homesteads were signals which even the least rationality could not fail to understand, and on every reappearance the natives found some favourite spot surrounded by new enclosures, and no longer theirs.' (West 1852, p. 272)

Some local historians have preferred to blame bushrangers: '… the atrocities and ill treatment dealt [to the natives] by escapees [made them] so incensed that they began to exact vengeance on all whites' (Historical Committee of the National Trust of Australia, Tasmania 1966, p. 96). The mainland Aborigine, Mosquito, was, somewhat unjustly, given the whole blame by Louisa Anne Meredith (1852). It seems likely that both the accumulation of atrocities from officials, sealers, whalers, settlers, convicts and bushrangers, and the explosion of enclosure in the 1820s precipitated the decision by a large proportion of the surviving indigenes to engage in war. One of their acts of war was to kill large numbers of sheep. Sheep appear not to have been regarded as a desirable source of food by the Aborigines (Robson 1983).

Early environmental impacts of sheep

The impact of the roaming flocks and herds on the native ecosystems seems to have been sharp and severe. The sediments and pollen in Camerons Lagoon on Liaweenee Moor, Central Plateau (Figure 1.2), a place with native vegetation still frequented by roaming flocks, give some intimation of this shock (Thomas and Hope 1994). One-sixth of the sediment accumulated over at least 8000 years filled the lagoon in the last 200 years, and the surrounding vegetation rapidly transformed from tussock grassland to the open shrubland with a short grassy understorey and much bare ground found today (Thomas and Hope 1994; Bridle and Kirkpatrick 1999). In Tasmania and other places in south-eastern Australia, the transition from Aboriginal to pastoral land use is marked by a huge spike of sedimentation in wetlands (Dodson and Mooney 2002, Leahy *et al.* 2005). Sedimentation declines to almost pre-European levels after the spike. Plants more palatable to sheep than to the native herbivores went into rapid decline, some are now extinct, and others survive only in places, such as cemeteries, roadsides and farmyards, where grazing is rare or absent (Kirkpatrick *et al.* 1988). Plants less palatable to the sheep than to the native animals, such as the heaths (Epacridaceae), increased in relative abundance. Invasive exotic plants were transported in the faeces and fleece of stock.

Two-legged dogs

In some places the roaming flocks were thought to be beset by predation from thylacine and devil (*Sarcophilus harrisii*), probably augmented by that of wild domestic dogs (*Canis lupus*) (Boyce 2006). In 1825, on Regents Plain on the Central Plateau, Captain Wood is reported to have replaced sheep with cattle because of predation of sheep by devils (Jetson 1989). In contrast, Morgan (1992) regarded the lack of predators in Tasmania to be a natural advantage, compared to the dingo-rich mainland of Australia, and quotes Widowson in 1829: 'the only dog to be feared is the two-legged one, and he cannot do much harm, unless your own shepherd be colleagued with him' (p. 57). How much the opinion of Widowson can be trusted is another matter. He was the agent for the Cressy Land Establishment, and there is a place on the Central Plateau called Wild Dog Plains, presumably not out of whimsy. Even if there were not wild dogs on Regents Plain in 1825, there would have been after 1830, when, in the face of the difficult demands of *The Dog Act 1830* many people released their dogs, which then formed packs that attacked both stock and native animals (Morgan 1992).

Two-legged dogs were undoubtedly active, as they are today. For example, between December 1816 and January 1817, 800 sheep were stolen from the Lake River area (Jetson 1989). The government regulatory response was to prohibit alteration of earmarks or brands (Jetson 1989). Nevertheless, 'The brand was obliterated, often with great ingenuity: the I became H, C was turned into G and P into B; the more daring blotted out all brands, by a heated shovel' (West 1852, p. 365). Some sheep duffers were hanged (West 1852). Large numbers of sheep continued to be stolen.

The pressure was not taken off the native predators. Thylacines were regarded as a major threat, to the degree that the Buckland and Spring Bay Extermination Association was formed in 1884 and the Spring Bay Council offered one pound per dead adult in 1888 (Lester 1994). The Association appeared to achieve its aims in the late 1930s when the last thylacine in captivity died, although there was a published reliable sighting, by National Parks zoologist Hans Naarding, as late as 1982, and unreliable sightings still abound. The thylacine is one of two animals that frequented sheep country that are thought to have become extinct, the other being the Tasmanian emu, probably hunted into extinction by the mid-nineteenth century.

Yeomen graziers

It was clear that, if the ongoing program to make Van Diemen's Land a major source of wool for English factories were to succeed, roaming flocks were not a solution. In 1820, a sheep census reported 182 468 individuals, considerably more than the 7400 non-indigenous humans recorded in 1821. The sheep were almost entirely crossbred, with Teeswater, Leicester and Bengal blood (Morgan 1992). Large, fenced, private properties run by respectable people, preferably with agricultural experience, might have been seen as the solution. They happened, in a rush, largely in the 1820s, when possessions worth fifty pounds, a good demeanour and persistence could gain eventual formal ownership of 640 acres of good land with a stream frontage, as well as access to convict labour. This rush set up the base for a pattern of society and land use that persists until today, down to the names of many of the wool-growing families. The social background of most of the people who took up their acreages in the period 1820–1831 was solidly middle class: '... retired officers like Captains Wood, Synott, Langdon and Fenton; merchants like Leake, Morrison, Bethune and Grant; professional men like Smith, Parramore and Gleadow; yeomen farmers such as Taylor, Archer and Lyne ...' (Reynolds 1969, pp. 61–62). The families of these settlers owned most of the 100 largest rural estates in 1875, when they were the most powerful group in Tasmanian politics (Reynolds 1969).

The future ruling class families of Tasmania did not take up their lands at an easy time. The Taylor family from Gateside, Scotland provide an illustration of these difficulties. They arrived at their grant, which they called 'Valleyfield', on the Macquarie River in summer 1823. In 1824 they were attacked by bushrangers, the Brady Gang, whom they fought off (Taylor 1985). In 1826, the hero of the fight against the Brady Gang, one of their six sons, George, was speared, one among approximately 170 European deaths in the Black War (Reynolds 2004). Since arriving at 'Valleyfield' in 1823, George had learned the local Aboriginal language and had mixed freely among the local indigenous people (Historical Committee of the National Trust of Australia, Tasmania 1966, p. 96). Familiarity was negated by the war for land, a war the Aboriginal people abandoned in the early 1830s for what they probably thought was a negotiated settlement (Reynolds 2004).

Bushrangers and blacks impelled many grantees to become absentee graziers, residing in the relative safety of the towns (Morgan 1992), while their sheep increased in numbers in the bush. Many of the large landowners took sheep 'on the thirds', a system whereby flock owners without land gave one-third of the increase in their flock for the privilege of using the land owned by others (Morgan 1992). The authorities did not like the 'on the thirds' system, because it facilitated sheep stealing (Morgan 1992).

Despite the killing of sheep and shepherds, and the use of fire as an economic weapon by the black warriors, the sheep numbers in Tasmania continued to rapidly increase (Figure 1.5). Some time between 1834 and 1839 the population attained one million (Davidson 1938). In 1854, one year after the cessation of transportation of convicts, the peak for the nineteenth century was attained, 1.9 million sheep (Davidson 1938).

A few of the yeomen immigrants of the 1820s were to lay the foundations for the wool industry in Tasmania, by importing and breeding sheep suited to purpose and environment. James Cox of 'Clarendon' has already gained mention. Others who established successful studs in this decade included William Archer, George Parramore, William Kermode and the Taylors (Morgan 1992). In the late 1820s the Van Diemen's Land Company spent thirty thousand pounds on Saxon Merinos and other

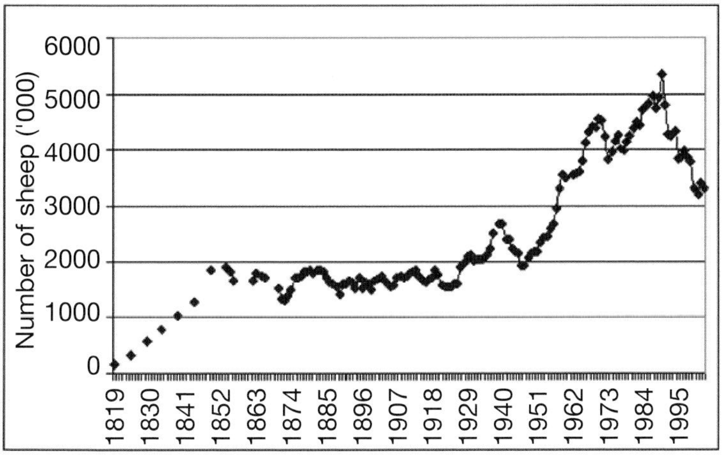

Figure 1.5. The growth and decline of sheep numbers in Tasmania.

high grade breeding stock (Massy 1990). During the early 1830s Thomas Henty imported 1500 Merinos of high quality to Van Diemen's Land, and the Forlonge family imported Saxon Merinos, carefully selected by Eliza Forlonge. By 1835 the family had 8000 sheep in Van Diemen's Land, half of which were sold to David Taylor of 'Winton' when the Forlonges moved from 'Kenilworth' to Victoria (Parsonson 1998). The success of the Tasmanian wool growers in the 1830s allowed the export of many thousands of sheep to populate the Victorian plains.

Figure 1.6. Hawthorn hedgerows established in early colonial times near Longford (Jamie Kirkpatrick).

Figure 1.7. Rock and paling fence from early colonial times (Jamie Kirkpatrick).

Breeding up flocks required that the stock be controlled through the establishment of barriers to movement. Hawthorn (*Crataegus monogyna*) hedges (Figure 1.6), post and rail fences, and barriers built out of rocks, palings and/or logs, were established widely in the gentler parts of the landscape between the 1820s and the 1850s. Louisa Anne Meredith (1852, p. 90), living in eastern Tasmania in the 1840s, commented that: 'Hawthorn hedges greeted me pleasantly again, with their old remembered verdure and fragrant blossoms; and those of gorse, the first I had seen since leaving England, would rival the growth of that sturdy mountaineer even on its native hills'. As well as the hawthorn hedgerows (Figure 1.6), some of the original fencing still survives (Figure 1.7). However, it seems that, until the 1850s, sheep still roamed over large unfenced parts of the countryside in the company of shepherds, whose main tasks were to contain their flocks within the property or leasehold and to prevent theft, rather than to provide the pastoral care characteristic of their role in Britain (Morgan 1992). In the 1850s fences patrolled by boundary riders, and fixed watering points, were adopted as a response to a sudden labour shortage (Cottle 1991). People left to dig for gold, at the same time as local political pressure had caused the cessation of transportation of convicts.

Another responsibility of the shepherds was to burn coarse vegetation. This ensured green pick for their flocks, and protected fences. Louisa Anne Meredith (1852, p. 109) noted that:

> 'Sheep owners know how serviceable occasional bush fires are, and generally arrange to burn portions of their sheep-runs at different times, so as to have new growth about every three years. Where this is neglected for a length of time, the rank luxuriance of the great brake fern and other uneatable plants, and the accumulated masses of dead wood, bark, and leaves, form such a body of fuel, that when a fire does reach it, the conflagration is thrice as mischievous in the destruction of fences than it would otherwise have been.'

Firing the landscape

Count Strzelecki (1845) expressed his strong disapproval of burning for green pick after visiting the Central Plateau, perhaps observing the soil erosion recorded in the sediments of Camerons Lagoon by Thomas and Hope (1994). Fire management techniques may have been picked up from the indigenous people, with whom Europeans had considerable congress, in both the intimate and broader senses, in the two decades after the British landing at Risdon Cove. Louisa Anne Meredith (1852, p. 108) was certainly of this opinion. She criticised Count Strzelecki for assuming that the colonists originated the practice, when the natives engaged in it frequently and extensively for very good reasons. Major Cameron, the late proprietor of a northern Midlands property, burned his native grasslands in patches when the breeze blew fresh (Figure 1.8). Major Cameron (personal communication 1998) believed that his management technique had been handed down from the earliest days of grazing in the valley. In the early nineteenth century, part of the same valley was held as freehold, possibly by Black Bill, one of the Tasmanian natives not on the indigenous side in the Black War. Black Bill was said to have been granted 100 acres as a reward (West 1852, p. 294), although Morgan (1992) suggests that he may have died before taking it up. In another part of the valley lay the Tasmanian domicile of Batman, an enthusiastic English warrior in the Black War. He received 2000 acres (West 1852, p. 294).

Certainly, there is sufficient historical evidence of people of Tasmanian Aboriginal descent surviving within the European society to make it likely that technology transfer persisted well beyond the Black War, a war in which such 'aborigines as were pacific' (West 1852, p. 296) were exempted from martial law along with the British. West (1852, p. 286) wrote of:

> 'Mrs. Dalrymple Brigge, a half-caste woman, (who) was rewarded with twenty acres of land for her heroism. She drew inside her house her wounded child, barricaded her door and fired through a crevice. The

Figure 1.8. Major Cameron, burning *Poa* tussocks, late 1990s (Simon Cameron).

> blacks attempted first to pull down her cottage, and then to destroy it by fire. The conflict lasted more than an hour, when relief came.'

On 26 May 1826, the Land Commissioner wrote in his journal when visiting Bagdad:

> '... from that we went to Kimberleys, a Native, his farm was originally granted in two locations to John Ingle ...' (McKay 1962).

The historical geographer, Simon Cubit, believes that a technology transfer from indigene to shepherd could also have taken place in the grassy high country, where patch burning took place as the snow melted (Cubit 1996). Cubit (personal communication 2005) writes of Bert Nichols, of reputed Tasmanian Aboriginal descent, who moved from his birthplace in the Campbell Town district to become a hunter and snarer in the high country in the early part of the twentieth century:

> 'Everywhere he went he burnt. In fact, the smoke from his fires was one way fellow hunters knew where he was. His fires were responsible for maintaining a wide variety of grasslands which he used to keep the country open and maintain high populations of native marsupials.'

Figure 1.9. Patch-burned shrubs in eroded grassy shrubland on the Central Plateau, late twentieth century (Jamie Kirkpatrick).

Burning of the native country, 'runs' as they are called in Tasmania, did not stop when labour became scarce in the 1850s (Figure 1.9). Mary Littlechild of 'Downward' described the way she fired runs on her property in the Midlands, near the Western Tiers, until the late twentieth century:

> '... we would ride our horses and you would get off, light a piece of bark and then ride along and light a fire. We were never frightened of fire. It generally happened before Christmas or a bit after because that time of year would give the best burn. You would ride along and there would be a big blaze behind you and you used to look round and say 'My word that's a beauty today. It won't go far, it will go to where it was burned last year or the year before.' (Gilfedder *et al.* 2003, pp. 11–12)

Jim McEwan of 'Trefusis' described his cessation of burning in 1967, due to the risk of legal liability. He used to burn about every seven years: '... mainly to keep the wattles down and to promote fresh growth, wethers would love the new growth on the burnt country' (Gilfedder *et al.* 2003, p. 8). Burning of run country as part of grazing management persisted widely in Tasmania until the late 1960s, and still continues on some properties (Gilfedder and Kirkpatrick 1998a; Leonard and Kirkpatrick 2004, Chapter 2).

Figure 1.10. Stump showing signs of past fires (Jamie Kirkpatrick).

Evidence of fire in the stumps of dry forest eucalypts, in the form of charcoal, occlusions and bleached rings (Figure 1.10), suggests that the fire regime between 1820 and 1850, in the Eastern Tiers and part of the Midlands, was very different from that experienced either before or after these dates (Julie von Platen, personal communication, 2006). Fires hot enough to leave signs in stumps decreased from about one per decade in the period 1740–1820, when Aboriginal people were the major occupants of all or most of the landscape, to about one per 30 years in the period from 1820 to 1850, then returned to about one per decade in the period 1850–1910 (Julie von Platen, personal communication, 2006). One possible explanation for the period of low incidence of evidence of fire in stumps is that the burning was more frequent and thorough, therefore less intense, when there were more people on the ground, so that insufficient fuel could accumulate to cause damage to trees when a fire was lit. This hypothesis assumes that Europeans lit many more fires in this back country in the period in which graziers had access to convict labour, than did the Aboriginal people in previous times. The three-year burning cycle suggested by Louisa Anne Meredith supports this hypothesis, unlike her description of fires in the bush in the 1840s at Riversdale, Great Swanport:

> 'By day, the effect of these great conflagrations was far from pleasant, causing an increase of heat in the air, and a thick haze over the landscape

> generally; whilst from the various point that the fires were raging, huge columns and clouds of dense smoke were seen rising, as if from volcanoes: but at night, the scene was often very grand; sometimes the fire might be watched, on any rising ground, spreading onwards and upwards, swifter and brighter as it continually gained strength, till the whole mountainside was blazing together; and after the first general flame had passcd away, and the great trunks of the trees alone remained burning, the effect resembled that of the scattered lights seen approaching a distant city at night. The rocky Schoutens glittered with partial lines and trains of fire, that marked their rugged and lofty outline like burnished gold amidst the darkness.' (Meredith 1852, p. 108)

An alternative, more likely, explanation for the decrease in evidence of fire in stumps is that the Aboriginal people were not lighting fires in the area in their normal manner, at first because they were fighting a war and did not want to reveal where they were, and later because the survivors of the war were on Flinders Island, not in the Eastern Tiers and Midlands. Accepting this explanation, the question then arises: what caused the increase in fire evidence in stumps after 1850?

A simple answer to this question might be that Europeans accessed the more remote country only after 1850. This seems unlikely. Some of the most remote alpine parts of the Central Plateau were occupied in 1840 (Cubit 1998). Graziers removed their stock from these areas by the late 1850s, not to return until the 1890s (Cubit 1998). If graziers were prepared to take stock to the highest parts of Tasmania, they could have been expected to take stock to the less lofty and inclement Eastern Tiers.

A change in climatic conditions, such that fire weather was markedly less frequent between 1820 and 1850 than before or after, is a hypothesis that is difficult to test, as climatic data are available from stations in south-eastern Tasmania only after 1855. However, the record from growth rings in celery top pines (*Phyllocladus aspleniifolius*) on Mt Maria does not suggest a period of unusually consistent moist conditions (Allan 1998). The 1850s had two severe drought years; the 1860s, 1870s and 1880s had one drought year each; but the period between 1855 and the late 1890s was, in general, one of normal rainfall (Figure 1.11).

Perhaps, on reaching numbers in excess of the long-term carrying capacity of the wooded run country, sheep changed the nature of fuel, or the nature of fuel accumulation, to produce dead and alive fuel assemblages that would burn intensely at shorter intervals than before. In one possible scenario, native grazing animal numbers would have been reduced to enable more sheep to be supported. A reduction in wallaby numbers would have reduced grazing pressure on shrubs, particularly heaths (Dickinson and Kirkpatrick 1986). When the sheep were rounded up to be taken elsewhere, whether to high country pastures or to other lowland pastures, native herbivore numbers would be too low to eat all available fodder, therefore allowing the build-up of litter to support fire. If the grasses cured before the return of sheep, and ignition

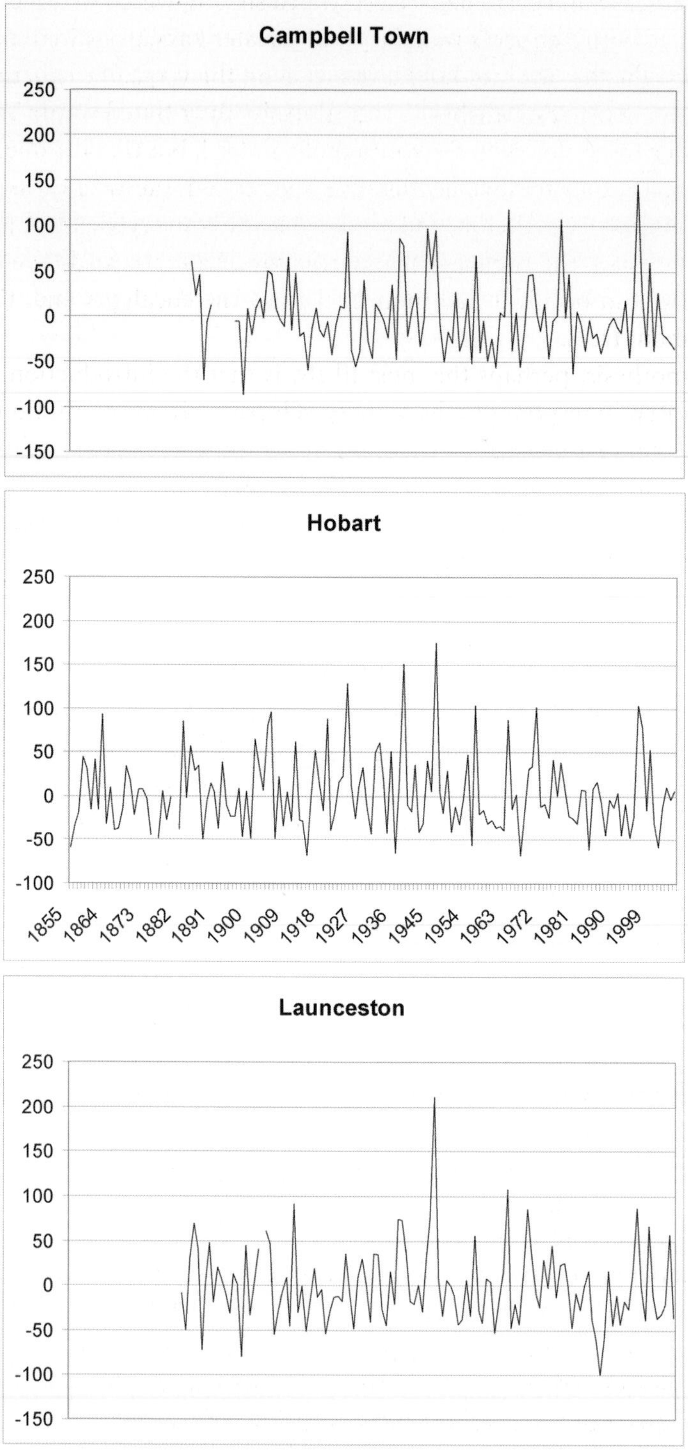

Figure 1.11. Percentage deviation of annual rainfall from the mean for three stations in eastern Tasmania (data from Bureau of Meteorology).

took place, the fire would carry between scattered shrubs, which would burn intensely. Low numbers of both Bennett's wallabies and forester kangaroos were noted by West (1852). He thought that forester kangaroos were on the verge of extinction. He wrote (p. 249) that the Bennett's wallaby '... is universally distributed across Tasmania, and in some locations were formerly very numerous indeed, but the war of extermination constantly waged at all seasons against this species, for the sake of its skin, has, in many places, entirely destroyed it, and rendered it scarce everywhere.' A third possibility that might explain the change is that overgrazing by sheep made fire less frequent. Thus, when fire did occur, it had more fuel from the eucalypts and, therefore, was more severe about trees.

A final hypothesis, perhaps the most likely, is that the introduction of fencing to leasehold country, in response to the scarcity of labour, after the cessation of transportation and the gold rushes, provided an incentive to burn to protect this investment. There was also a dramatic expansion in the leasehold area between 1847 and 1853, from 1.1 million acres to 2.3 million acres (Davidson 1938). Much of the Eastern Tiers may have been first leased at this time.

Rabbits, scab and other pestilences

It is common for the population of a newly introduced organism to grow exponentially, then to fall sharply after attaining its peak, usually as a result of the self-induced degradation of its food supply or because of density-dependent disease. The post-peak fall of sheep numbers in Tasmania was not great, from 1.9 million sheep in 1854 to an average of 1.67 million sheep between 1859 and 1924 (Figure 1.5, Davidson 1938). The lowest population in this latter period was 1.5 million in 1869 (Davidson 1938). This low point was thought by Davidson (1938, p. 345) to be:

> '... associated with (a) the prevalence of sheep scab (a disease caused by a mange mite), which necessitated the passing of the 'Scab Act of 1870'; (b) the prevalence of fluke in certain pasture areas; (c) the development of the rabbit pest, which necessitated the passing in 1871 of 'An Act to Provide for Destruction of Rabbits in Tasmania'; (d) the persistent fall in the price of wool which dropped from 22d. a pound to 15d. a pound between 1862 and 1870.'

Changes in fire regimes and problems with rabbits (*Oryctolagus cuniculus*) and diseases seem likely to have been different ecological reflections of the same phenomenon – full occupation of the native pastures of Tasmania by sheep.

Wild grey rabbits were first introduced to Tasmania by William Crowther in the early 1820s. Within a few years there was a report in the *Colonial Times and Tasmanian Advertiser* (11/5/1827) of thousands of wild rabbits on some estates (Jetson 1989). Yet, a serious rabbit problem was only perceived in the lowland sheep-grazing country in the latter half of the nineteenth century, and on the Central Plateau in the 1920s (Scott

Figure 1.12. Lawn in Midlands created by grazing of sheep and wild animals. A white gum (*Eucalyptus viminalis*) woodland with a sagg (*Lomandra longifolia*) understorey is in the middleground (Jamie Kirkpatrick).

1955; Shepherd 1973). Part of an explanation for this slow development of a problem on the Central Plateau may lie in the tendency of rabbits to prefer to graze on short lawns. The grazing of kangaroos, wombats and wallabies can create lawns, but these tend to be concentrated in the most fertile and moist places. The normal outcome of heavy grazing by sheep is a lawn well suited to rabbits (Figure 1.12). Sheep may have taken longer to create lawns in the high country than in the dry lowlands because stocking rates were lower and shrubs more abundant. Isolated rabbit problems did occur in early-settled districts, well before sheep numbers attained their maximum in the nineteenth century. For example, Louisa Anne Meredith (1852, p. 96) described the wild rabbit problem near Swansea in the 1840s. She also described the current wild rabbit control technique:

> '... their abundance all over the district is now so great as to sometimes to be troublesome and mischievous. The fence-banks overgrown with huge gorse-bushes form admirable retreats for the pretty little animals, and they increase so rapidly, and make such depredations among the young turnips and springing corn, that occasionally a proclamation of 'war to the knife' goes forth, when a group of sportsmen and spaniels take the field; at such times the golden-blossomed gorse-bushes do not escape, many of the

> most effectual 'covers' being burned down in order to dislodge the rabbits.'

The sandy banks that are so widespread throughout the Tasmanian sheep-grazing country provided an ideal situation for rabbit warrens. Responses to the rabbit problem were various. In the 1870s John Taylor of 'Winton' wrote (Taylor 1985, p. 151):

> 'During the 1870s the rabbits (were) very troublesome and caused a great loss. I am obliged to enclose the sheep runs with paling fences. Seven miles of paling fences we put up in 1874. I have to keep a lot of men catching rabbits. I give them 3/- a dozen and they carve great wages at that price. An example of their number … five of us shot 700 in one day.'

Paling fences could only be effective exclusion devices if they were attached to buried netting. Partly buried wire mesh netting became the fence of choice to exclude rabbits, but was too expensive to be widely used outside the best country, and still required frequent patrol to detect excavations by wombats and echidnas (*Tachyglossus aculeatus*).

Shooting and exclusion were fates shared by most native mammals, although it takes more than a wire mesh fence to exclude a brushtail possum. Native animals also shared the impacts of other rabbit control mechanisms. Steel-jawed rabbit traps were set in lines near rabbit burrows, and concealed with soil. Barney Gatenby recalled catching the occasional bandicoot (*Perameles gunnii* and *Isoodon obesulus*) and native hen (*Gallinula mortierii*) in the 12 traps he had as a child in the Midlands (Gilfedder *et al.* 2003). Cats (*Felis silvestris catus*) were caught, and were also killed by the packs of dogs used to rouse rabbits and other game for shooting (Gilfedder *et al.* 2003).

Rabbits could die by inhaling poison gas in their warrens or, more commonly, by ingesting poisoned baits. As early as 1871, John Bisdee 'boiled strychnine crystals in muriatic acid, added the mixture to boiling water before pouring it over crushed oats or wheat' (Jetson 1989, p. 122). He then scattered the bait in places frequented by rabbits. Phosphorus and arsenic were other popular poisons, to be followed by sodium fluoroacetate (1080) in the 1950s and later. Barney Gatenby recalled:

> 'Before the '30s they used a poison cart with phosphorus poison. They laid a furrow and dropped the poison in the furrow. The next poison they used was strychnine and the furrow had to be free-fed with carrots or apples for three days and then poisoned on the fourth day. … but the trouble was as the rabbits took the bait in their last dying hours they used to squeal and that would frighten quite a number away. … It was nothing to pick up 500 to 1000 rabbits in a furrow …' (Gilfedder *et al.* 2003, p. 3)

Primary and secondary poisoning of non-target species was inevitable. Any native animal that ate the bait, or the victims of the bait, died.

By the 1960s rabbits were no longer a major problem, because of the effectiveness of myxomatosis, the rabbit disease, in destroying populations of the animal. The introduction of myxomatosis was not popular with some sections of the community due to the suffering it caused. It is distressing to watch the slow death of an animal that is widely reputed to deliver Easter eggs. Spike Milligan (quoted in Jetson 1989, p. 123) engaged in poesy on the subject:

> 'A baby rabbit
> With eyes full of pus
> Is the work of scientific
> Us.'

But, at least the dying rabbits are too debilitated to squeal.

There were other reasons not to want myxomatosis. Rabbits were a source of easily available free meat for the rural poor. There was also a substantial market for rabbit skins at prices that meant that rabbiters could gain a respectable income. For example, rabbit skins returned three to six pence in 1900 (Prevost 1988). Many people had an incentive to avoid wiping out rabbits. Bob Green reminisced on the introduction of myxomatosis around Oatlands (Gilfedder *et al.* 2003, p. 39):

> 'There were one or two farmers who got into trouble because they decided it was hopeless to try and grow sheep, so they thought just get rid of the sheep and use the rabbits. By law they were supposed to poison and one thing or another and they didn't. They were sort of saving them until the wintertime when the skins were of the best quality. When the Ag. Department were there at Oatlands with these infected rabbits, there were half a dozen of the old diehard rabbiters going crook …'

Myxomatosis resulted in a marked increase in the carrying capacity of many pastures in the Midlands. One property that had 5500 sheep was able to support 2000 more (Prevost 1988). White clover was no longer chewed into invisibility.

Oral histories from the Midlands give the strong impression of rapid change in the species composition and relative abundance of the mammal fauna around the time of the introduction of myxomatosis (Campbell Town District High School 2002; Gilfedder *et al.* 2003). These included decreases in bandicoot and hare (*Lepus europaeus*) numbers, and increases in cat, deer (*Dama dama*), wallaby, brushtail possum and Tasmanian devil populations. There is no equivalent information on the effect of reduction of rabbit numbers on native plant species, but it can be deduced from data from mainland Australia that tree establishment became more likely in the drier areas and that many native herbs and grasses would have increased in abundance (Lange and Graham 1983; Leigh *et al.* 1987; Tiver and Andrew 1997).

Bob Green, a zoologist who was brought up in the Midlands, has suggested that prey substitution led to the declines in bandicoot and hare numbers (Gilfedder *et al.* 2003, p. 38):

> 'The first year after myxo went through and made such a mess of the rabbits, when spotlighting, we would think 'Ah, there's a rabbit' and it

> would be another bandicoot. But within 12 months all the bandicoots were gone. And that was just predator pressure created by the lack of rabbits as food ...'

The animals that were observed to increase were those likely to take poisoned baits or be subject to secondary poisoning.

Before 1850, when their numbers were rapidly growing, sheep, although not always of the finest breeds for wool production, were generally fecund, fit and healthy. The transmission of disease was limited in the broadacre pastures they frequented. The green sheep blowfly (*Lucilia cuprina*), which is the main cause of strike, was yet to be introduced. Even scab, an itching disease caused by the mange mite (*Psoroptes ovis*), is thought to have been moderated in its impact by an excellent diet (Morgan 1992). Erroneous theories of the cause of scab included lack of shearing, as suggested by Edmund Curr, and, puncturing by the seeds of native grass, as posited by Widowson (Morgan 1992).

Scab was a problem because sheep rubbed themselves to relieve their itching and thereby damaged their wool. A dip in an arsenic or tobacco solution was followed by washing in fresh water and brushing off the scabs. Sheep became infested with mange mite through contact with infested sheep. The elimination of the disease thus required concerted community action. There always being someone not prepared to work for the broader good, the *Scab Act 1870* was passed. Inspectors of Stock were so successful in enforcing the control of the mite in the declared Sheep Districts that, in 1894, the Chief Inspector of Stock declared the State scab-free (Jetson 1989).

Liver fluke (*Fasciola hepatica*) was, and is, a more ecologically specific disease. It has a life cycle that involves the livers of aquatic snails that live in swampy areas, a free swimming form, and the bile ducts of the liver of any mammal unfortunate enough to ingest its cysts (Cottle 1991). Sheep have little inherent resistance to liver fluke compared to other animals, such as cattle. Infested sheep can suffer fatal bleeding, reduced growth, anaemia and bottle jaw (Cottle 1991). Much of the naturally grassy country of Tasmania lay in broad valleys in which swamps were common. These swamps would dry during drought, but remain green, attracting sheep from the parched upslope pastures, to ingest heavy loads of liver fluke cysts. In these circumstances mortality can be high, with one-fifth of one flock lost on the Central Plateau as late as 1971 (Simon Cubit, personal communication). Simon Cubit has suggested liver fluke as one possible cause of the retraction of transhumant sheep grazing from the western part of the Central Plateau in the late 1850s (Cubit 1998). Climatic data from Hobart shows a dry period in the 1850s, consistent with this hypothesis.

There was no chemical control for liver fluke before 1916, when carbon tetrachloride drenches began to be used, not fully effectually, for this purpose (Cubit, personal communication). The earlier solutions for the liver fluke problem were to drain the swamps, or to separate sheep from the swamps through fencing. Early nineteenth century drainage works can be seen in many of the valleys of grassy Tasmania (see Figure 1.13). Willows (*Salix* spp.) were also used to help dry wet places. They spread

Figure 1.13. Convict era drains and willow plantings in native pasture near Bronte Park (Jamie Kirkpatrick).

from these places, and other points of introduction, along most of the major lowland streams. Even a broken bit of willow branch is capable of forming a new tree. Many are deposited in sediments as a result of floods. The full ecological consequences of wetland drainage and willow invasion will never be known. However, very few lowland wetlands on fertile ground survived until the late twentieth century (Kirkpatrick and Harwood 1983a,b). Willows displaced native riparian vegetation, likely to have been rich in currently rare or threatened plant species (Daley and Kirkpatrick 2005). Pink willow root systems also displaced the native aquatic plants of streams (Chappell 2000). These changes, in turn, influenced the nature of the aquatic fauna.

Another sheep disease related to wetness is footrot. This bacterial infection of hooves is enabled by the skin damage that occurs in wet tissue. This disease and fly strike are associated with areas around watercourses, rather than open, sparsely grassed plains and hilly country (Cottle 1991). Although native blowflies can lay their eggs on sheep, problems with blowfly strike largely relate to an exotic blowfly from Asia and Africa, the green fly. The blowflies pupate in the soil and emerge to lay eggs, which turn into maggots, on the skin of the sheep.

Like the rabbit, the introduced blowfly was present well before it became a major management problem for wool producers. The first recorded green fly strike on a sheep in Australia was in 1880 (Massy 1990). By 1910 blowflies were seen as a major problem in eastern Australia, although they did not appear to be a problem in Tasmania until the late 1940s (Massy 1990). Massy (1990) suggests that the species may have evolved into a particularly virulent form, in an Australia rich in sheep, both dead and alive, many of which had wrinkled skins during the time in which the green fly spread from its initial introduction points. However, Cottle (1991) suggested that carcases may not

Figure 1.14. The Tasmanian Merino before the Vermont phase (Gibson 1887).

be important in the reproduction of the introduced blowfly as they are colonised by other blowfly species with larger, more competitive maggots.

Breeding for environment, purpose and fashion

There is some capacity to breed attributes into sheep that increase their disease resistance, either by culling those that appear particularly susceptible or by breeding for skin and wool characteristics that do not encourage particular diseases. Such selection has to be balanced by the promotion or retention of characteristics that contribute to the commercial viability of the animal. Thus, sheep breeding is a highly skilled process that, ideally, results in animals that return the best profit to a wool grower, by maximising the difference between costs and returns. The animal needs to fit the environmental and economic conditions that prevail on a particular property within a particular state and country at a particular time. By 1850, Tasmania was the major centre of sheep breeding in Australia (Day and Jessup 1984). Sheep studs developed animals with considerable Saxon Merino blood that produced more and more wool per individual and that were well adapted to local conditions (Figure 1.14). Stud rams from a large number of well-established properties formed a profitable export to the mainland of Australia (Massy 1990). An outstanding enterprise was that of the Gibsons at 'Scone' (Figure 1.16).

In his tome on the Australian Merino, Massy (1990, p. 418) waxed lyrical on the subject of the product (Figure 1.14) of nineteenth-century breeding in Tasmania:

> '... the old plain-bodied, long-stapled, fine-woolled Tasmanians of the late last century were extraordinary sheep. With their German-Negretti stamped productivity and modernity of outlook, and fine frame, they were

Figure 1.15. Wrinkled Merino ram with Vermont blood (Gibson 1887).

> probably the pinnacle of fine-woolled Merino breeding, until Sir Walter Merriman changed the rules and successfully combined a Peppin carcase and Saxon wool.

Massy (1990, p. 418) regretted the later introduction of genes from an American sheep, the Vermont:

> 'Sadly, the Vermont fashion (either directly or by imitation) ruined the old Tasmanian type, which has never since been recaptured. The tragedy is that it was none other than William Gibson of Scone in the early 1880s who began the move in Tasmania to experiment with the American sheep, and so, because of his influence, he helped sow the seeds of destruction.' (Massy 1990, p. 418)

The Vermont-influenced Merinos (Figure 1.15) had a highly wrinkly skin, contrasting markedly with the smooth skin of the older Tasmanian sheep. The commercial basis for breeding for wrinkles would have been an increase in wool production per sheep. Greater skin area equalled greater wool volume. Massy (1990, p. 504) argued that the development of a fashion culture destroyed the utility of the wrinkly sheep:

> 'The trend towards more and more, and tighter and tighter, wrinkle can be seen in sequential photographs of show sheep of the time, until by the

Figure 1.16. 'Scone' in the late nineteenth century (top) (Gibson 1887) and in the twenty-first century (bottom) (Kerry Bridle).

Figure 1.17. A contemporary Merino ram (Diana Cameron).

> 1900s the pursuit of corrugated-iron sheep had stripped the animals of any redeeming features. They had become smaller; were narrow-chested; had shocking conformational faults such as bad pasterns, devil's grips, sloping rumps and virtually every bad body fault imaginable. Their fleeces exhibited serious unevenness, declining quality and hair, not to mention excessive grease and low yield. And they were losing fertility rapidly, a factor compounded by their deep flank and skin folds which made it hard for lambs to find teats. They were in every way a disaster.'

He also stated that the Vermont sheep had very poor constitutions, did not travel well, were subject to blinding by grass seeds, and suffered high mortalities in drought (Massy 1990). After the great drought of the 1890s and its breaking in the early twentieth century, wrinkled sheep disappeared from Australian flocks. If they had not disappeared, they would probably have been abandoned because of the imminent ubiquity of blowfly strike, to which wrinkled sheep are highly susceptible.

In the latter half of the nineteenth century wool growing on the Australian mainland was extending into country with dramatically more severe environmental conditions than the wool-growing areas of Tasmania. The lack of fitness of the Vermont-influenced Merino sheep that emerged from the still prosperous Tasmanian studs in the last two decades of the century helped impel the mainland graziers to breed types of sheep better suited to their conditions. Fifty Tasmanian studs were selling in Sydney in

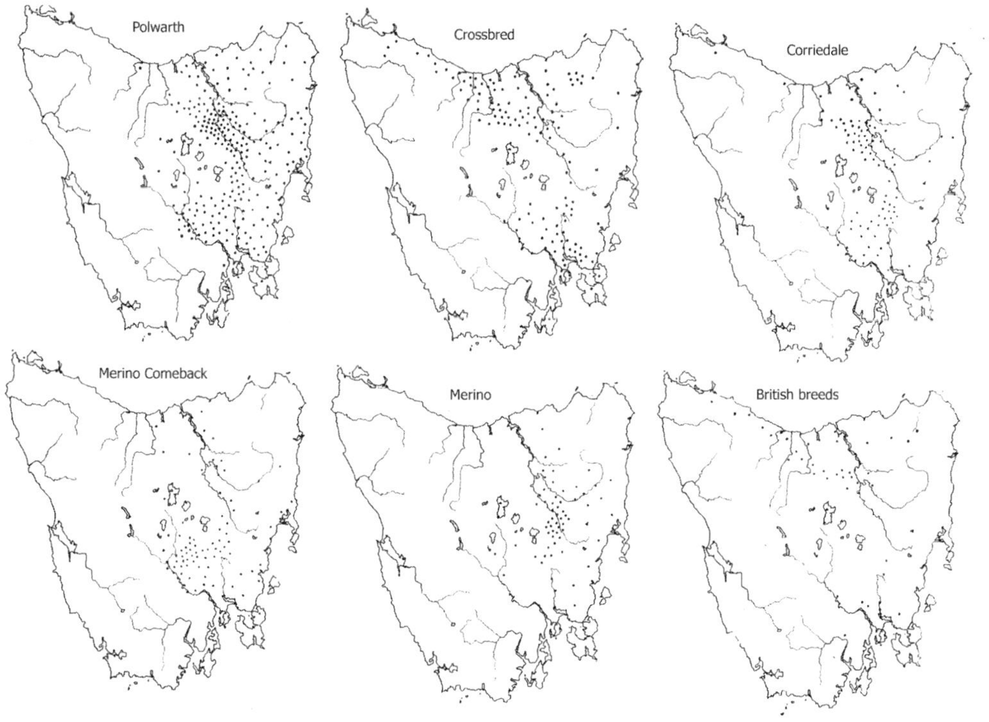

Figure 1.18. Distribution of major sheep breeds in Tasmania in the 1960s (Scott 1965).

the late 1890s. Soon after the burst of the Vermont bubble, no Merino studs were registered in Tasmania (Massy 1990). However, the Taylors at 'Winton', and others, continued their Saxon Merino breeding program, gradually de-wrinkling their sheep (Figure 1.17). These superfine-wool producing sheep are still very fit for particular environments and markets. They are resistant to skin diseases and fleece rot, and, from meagre forage in run country, produce superfine wool of high tensile strength. Some properties with relatively scrubby bush introduced Peppin blood to develop a solid fleece resistant to incursion by twigs (Prevost 1988).

During the twentieth century, the Merino was partly replaced in Tasmania by the Polwarth sheep, a comeback based on Merinos. This breed constituted almost half of the Tasmanian population by 1980, with Merinos declining to less than 10 per cent from 50 per cent in 1920 (Massy 1990). Polwarth sheep proved particularly suitable for the rougher run country, with their open faces suited for access to rough forage and their tough constitutions that made them good travellers. They also withstood fleece rot and did not easily cast (Massy 1990). In the mid 1960s they were widespread, but most abundant in the northern Midlands, while Merino comebacks were concentrated in the western Midlands around Bothwell, Hamilton and Ouse and Merinos in the southern Midlands between Campbell Town and Tunbridge (see Figure 1.18, Scott 1965). However, in the late twentieth century and early twenty-first century Saxon Merinos have resurged in popularity in the run country (Chapter 2).

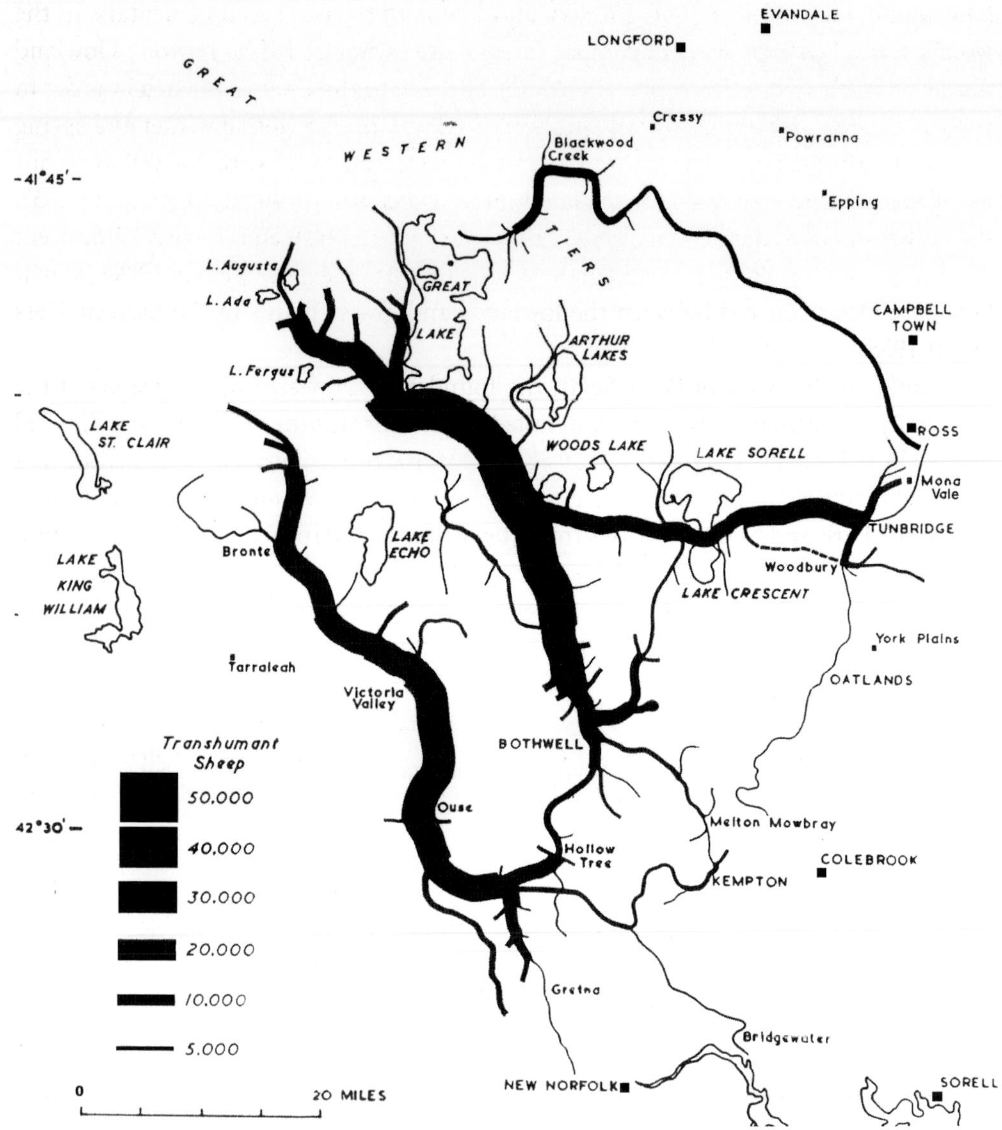

Figure 1.19. Transhumance routes in Tasmania in the early 1950s (Scott 1955).

Transhumance

As the numbers of sheep and Britons continued their rapid increase in Tasmania in the first half of the nineteenth century, the occupied grazing land reached its carrying capacity, compelling a search for more marginal lands suited to sheep. By 1840 much of the better grazing land below 1000 m had been alienated. At least two graziers had their stock on the higher grassy country of the Central Plateau at this time (Calder 1840). An explosion of leases on this country followed Calder's initial survey. Graziers had rapidly realised that high country and low country were complementary in the wool-growing enterprise. For example, in 1847, the owner of 'Mt Morriston', a lowland property south-east of Ross, sent thousands of sheep to the Central Plateau in order to conserve the fodder on his lowland property for lamb production in winter and spring (Crawford 1966). This was a common stratagem (Jetson 1989) with 350 000 transhumant sheep being estimated for the late nineteenth century (Shepherd 1973). In 1857, the Government Statistician estimated that sheep on the plateau returned 25 per cent per annum on the capital invested in the wool-growing enterprise (Jetson 1989). Transhumance also occurred between the lowlands and grassy plains in the Eastern Tiers (Scott 1955).

Thanks to the work of Peter Scott, the foundation Professor of Geography at the University of Tasmania, we have a detailed picture of transhumance on the Central Plateau in 1953 (Scott 1955; Figure 1.19 above). At this time, the graziers in the Bothwell area sent 60 000 sheep up the Lake Highway. Forty-five thousand sheep moved up to the Bronte Park and higher pastures from the western Midlands, centred on Hamilton. Twenty-five thousand more sheep were herded along the Interlaken Road from Tunbridge, to ultimately graze in the country around Great Lake, while approximately 2000 sheep moved to the high country north of Great Lake from the Ross and Blackwood Creek areas. In toto, this constitutes a much smaller number, approximately 135 000, than the 200 000 transhumant sheep recorded in 1933 (Shepherd 1973). Scott (1955) attributed the decline in transhumant sheep numbers to land deterioration through burning and overstocking, rabbits, a decreasing availability of shepherds and the effects of hydro-electric development. He predicted a slow continuing decline in transhumance, but doubted that it would completely cease:

> '... if only because transhumant sheep furnish high quality wool .... Nor is there much sign of a better use being made of the lake runs which alone could ensure them a permanent place in Tasmania's pastoral economy.' (Scott 1955, p. 172)

Consistent with this prediction, the number of transhumant sheep had declined to 91 000 by 1971, with most of the decrease occurring on land above 1000 m (Shepherd 1973). Shepherd (1973) attributed this decline to deteriorations in pasture quality and profitability, a lack of labour and the increased level of improvement in lowland and low plateau pastures rendering transhumance unnecessary. The return on capital investment on the upper plateau was said to be 'only' 7.6 per cent per annum (Shepherd 1973). Erosion and unpalatable shrub invasion were rife (see Figure 1.9; Mitchell 1962; Shepherd 1973).

By 1972, when the Royal Society held a symposium on the subject of the Central Plateau (Banks 1973), there was a strong scientific consensus that sheep grazing was incompatible with the maximisation of water yield and the maintenance of nature conservation values, by this time regarded as 'better uses' on the higher parts of the plateau, even by those with a production focus (Yates 1973). While there were a few leasehold areas, such as the Downies', that had not been burned by shepherds, and, as a consequence, had little bare ground and few kerosene bushes (*Ozothamnus hookeri*), the prevailing management regime consisted of patch burning followed by stock grazing. Burning produced green pick and temporarily set back the unpalatable shrubs, but also resulted in accelerated erosion and the establishment of new shrubs. The old shrubs simply re-sprouted. As Bill Jackson (1973) quipped, the short-term cure was the long-term disease. Bill was a bluff, bearded botanist of Tasmanian birth whose favourite recreation was trout fishing on the Central Plateau. He had had plenty of opportunity to observe the deterioration of the landscape. Two postgraduate students under his supervision, Bert Shepherd (1973) and Ivo Edwards (1973) provided hard evidence of the negative effect on water yield caused by the prevailing pastoral management regime. This negative effect was largely because the alpine shrubs stripped fog, and the eroded soils did not release water in a steady flow, but rather in flashes, after heavy rain or at snow melt. Lots of water flowed off the land, then uselessly to the sea, instead of slowly seeping through thick vegetation and soils into dams built for hydro-electric power production.

In 1965, the Lands Department had initiated a system of annual grazing licences for the high country Crown Land, replacing the longer leasehold agreements of the past. According to Shepherd (1973, pp. 171–172):

> '... the Department renews temporary licenses without prior inspection of the land under lease ... . The Department does not even regulate the numbers of stock using the region. ... The issuing of temporary licenses has led to mismanagement by the lessees as well. Such short-term land use offers no incentives for graziers to erect new fences, or even maintain existing ones, or otherwise manage the land for its long-term benefit. Since no action is taken against stock straying and grazing unleased Crown land, the grazier has been given free rein over a valuable catchment area by measures introduced to increase Departmental control for enhancement of catchment values.'

The author of these words was later to be employed by the Lands Department which, under its conservation-minded director, Doug Doyle, was to set up the Central Plateau Protected Area, which included most land above 1000 m. Supervision of grazing licences was tightened considerably, and burning, whether by shepherds, fishermen or shooters, largely prevented, at least within the Protected Area, by the assiduous ranger, Val Dell. Patch burning continued on the Liaweenee Moor and other freehold areas.

In the early 1980s the eastern Central Plateau was shown to have a high concentration of plant species with highly restricted distributions (Kirkpatrick and Brown 1984). These local endemics were concentrated in the very same high altitude areas that were grazed by sheep. In the latter part of the decade Gibson and Kirkpatrick (1989) provided evidence of a critical threshold for grazing impacts on the native vegetation of the

Central Plateau at 1000 m. Below this elevation, stock grazing in native pastures increased the number of native species in the vegetation, and there was little or no erosion. Above this elevation, stock grazing appeared to have negative effects on vegetation and soils. Later work on the Liaweenee Moor showed that bare ground in more sheltered areas would revegetate slowly after fire at a decade's scale, even with continued sheep grazing (Bridle *et al.* 2001), but not in the more wind-exposed situations, where vegetation cover increased at about 1 per cent per annum where sheep were excluded, but decreased where they continued to graze (see Figure 1.2, Bridle and Kirkpatrick 1999).

The eastern Central Plateau centre of endemism and interesting glacial outwash features provided a justification for including much of this area in the expansion of the Tasmanian Wilderness World Heritage Area in 1989. After a brief period of argument, no more grazing licences were issued. The arguments for persisting with grazing were partly cultural and partly ecological. Transhumant grazing on the high plateau had been a major part of the life of many Tasmanian families for a century and a half. They had built huts and fences that reinforced their sense of place and were part of a living history, a history that now seemed to be belittled and even denied. The movement of large flocks of sheep to and from the plateau along the stock routes (see Figure 1.19) was a twice annual event, celebrated by more people than those directly involved in the activity.

Those graziers who did not use fire as a management tool, and therefore had runs in good condition, felt especially aggrieved to lose their heritage. They, and others, put forward the ecological arguments expressed earlier by the agricultural scientist, Yates (1973, p. 156): '… a general ban on grazing … may not provide … in the long term perhaps the situation easiest to manage from the point of view of fire and rabbit control'. The logic of the fire argument was that the removal of sheep would lead to increased fine fuel levels and therefore to more severe and extensive fires than would otherwise be the case. However, in the decade and a half since the cessation of sheep grazing in the World Heritage Area marsupial grazing has kept fine fuels low and ignitions have been few, with only very small areas of alpine pastures being burned. The logic of the rabbit argument was that the shepherds were effective in keeping rabbit numbers under control. However, the outcome of heavier marsupial grazing does not appear to have favoured rabbits to the same degree as the outcome of sheep grazing, and rabbits, while still present, have not been in high numbers (Bridle *et al.* 2001).

Transhumance continues to places like Liaweenee Moor and the plains at lower altitude, but is a dun shadow of its former self. Sheep are now absent from most of the high plateau, permitting an extremely slow recovery of vegetation and soils. Stock routes are only lightly used. This has had some negative ecological consequences. For example, rare plant species that need very open conditions have disappeared from the Interlaken stock route between Tunbridge and the Western Tiers (Louise Gilfedder, personal communication, 2005).

Technology

Technological developments have been critical in influencing the relationships between people, sheep and nature in Tasmania, although, in some cases, these have largely had the effect of increasing profitability, without affecting nature. Such a development was the shearing machine, first adopted in 1890 at 'Egleston', a property supporting 40 000 sheep (Robson 1991). By 1914 there were 337 properties with these machines in Tasmania (Robson 1991).

In 1914, when young Australian men began to die en masse in the trenches of Gallipoli, France and Belgium, the wool-growing pastures of Tasmania were in a poor state (Robson 1991). White clover (*Trifolium repens*) and cocksfoot (*Dactylis glomerata*) were expensive to sow and likely to disappear rapidly under rabbit grazing pressure. The soils that supported the pastures were often so poor in nutrients that sheep ate bones, ash and corpses of animals to rectify deficiencies in their diet (Robson 1991).

The potential value of artificial fertilisers for increasing wool production was realised very early. In 1855, Theodore Bartley wrote:

> 'In the course of the last year, a cargo of Peruvian guano was brought to Hobart Town … . Of this cargo I purchased about eight tons … and have used at the rate of 200 lbs. to the acre, sown broadcast by hand … . The effect on the pasture is really extraordinary.' (Phillips 1987, p. 64)

His efforts in publicising the use of fertilisers (Bartley 1855) seem to have had little effect on sheep numbers, perhaps indicating that the economics of guano use were unfavourable for this purpose.

The fertiliser revolution began in the 1920s. Joan Prevost (1988, p. 42) wrote of 'Clyne Vale':

> 'When subterranean clover was introduced in about 1928 the increase in stocking rates was phenomenal. It doubled stock numbers, plus increasing fertility, by the deposit of nitrogen, formed on the leguminous root system. Together with the use of artificial fertilizers available, made from rock phosphate and the formula treatment at Risdon in Hobart, pastures became capable of carrying vastly increased stock numbers.'

Top-dressing of pastures took off from 1926 onwards, with the reorganised Department of Agriculture encouraging this as one of its major goals (Kellaway 1989). Between 1923/24 and 1936/37 the area of fertilised pastures increased from 52 077 to 191 928 acres (Davidson 1938). Between the mid 1920s and 1936/37 the amount of fertiliser used on pastures increased from an estimated 250 tons to a reality of 10 000 tons, enough for almost 20 000 acres (Kellaway 1989). In the same period, sheep numbers increased from 1.6 million to 2.2 million (Davidson 1938). Most of the new sheep were destined for the table, rather than the shears, as pasture improvement was directed largely towards producing prime lambs on previously cropped paddocks that had a vegetation cover composed of 'sorrell, Yorkshire fog, browntop, Kangaroo grass, danthonia and native weeds' (Kellaway 1989, p. 236) before fertilisation. By the mid 1930s pasture improvement on wool-growing properties was largely confined to the paddocks used for breeding up sheep, or for fodder production (Kellaway 1989). The application of superphosphate to run country was just starting.

The late 1930s were the beginning of a massive transformation from native grassland and grassy woodland to improved pastures in the wool-growing areas of Tasmania, a change that allowed the number of sheep to increase to its all-time peak of 5.3 million in 1990.

The area of artificially grassed land in Tasmania in 1933/34 was 671 728 acres (Lowndes and Maze 1937). By 1955/56 improved pasture covered one million acres, increasing to 1.84 million acres in 1966/67 (Solomon 1972). Barney Gatenby described the process of improvement on 'Bicton'.

> 'We started developing the lower country in the late '50s and we had done most of what we have done now by 1975. It involved getting a bulldozer in and getting all the dead stumps up, getting the gorse patches up and clearing and burning. ... we wanted to get over the country and get it laid down with clover and grasses ... sub-clover was our thing that did the most for us, with the superphosphate of course. ... It has only been turned over once, put into turnips and molybdate super in the first year, then laid it down in the next.' (Gilfedder *et al.* 2003, p. 6)

In the late 1940s aerial application of superphosphate was initiated in Tasmania, on 'Beaufront', by the von Bibra family. Kenneth von Bibra described some of its effects:

> 'It certainly had an impact on species such as kangaroo grass ... . The blue tussock had been given a hiding by the rabbits but that was still abundant and still is in places. There was a native clover and once that was given the advantage of super, it reacted very strongly, although after the introduction of subterranean clover it didn't seem to cope so well with the competition.' (Gilfedder *et al.* 2003, p. 26)

The deep-rooted perennial kangaroo grass (*Themeda triandra*) has a well-known aversion to both ploughing and fertilisation (Kirkpatrick, Gilfedder *et al.* 2005). The native clovers may have been *Glycine* or *Desmodium* species. Unploughed but fertilised paddocks were often top sown with introduced pasture species, maintaining the benefits of shelter from the native tussocks with increased fodder production.

The discovery of the lack of certain trace elements in poorer Tasmanian soils allowed the development of improved pastures in places, such as Epping Forest, that previously only supported gravel and firewood extraction. From 1972 onwards the availability of royalties for trees used for woodchips helped finance the conversion of native forest to pasture (Prevost 1988). Much of Epping Forest disappeared in the 1970s, as did much of the other bush in the Tasmanian Midlands (see Figure 1.20,

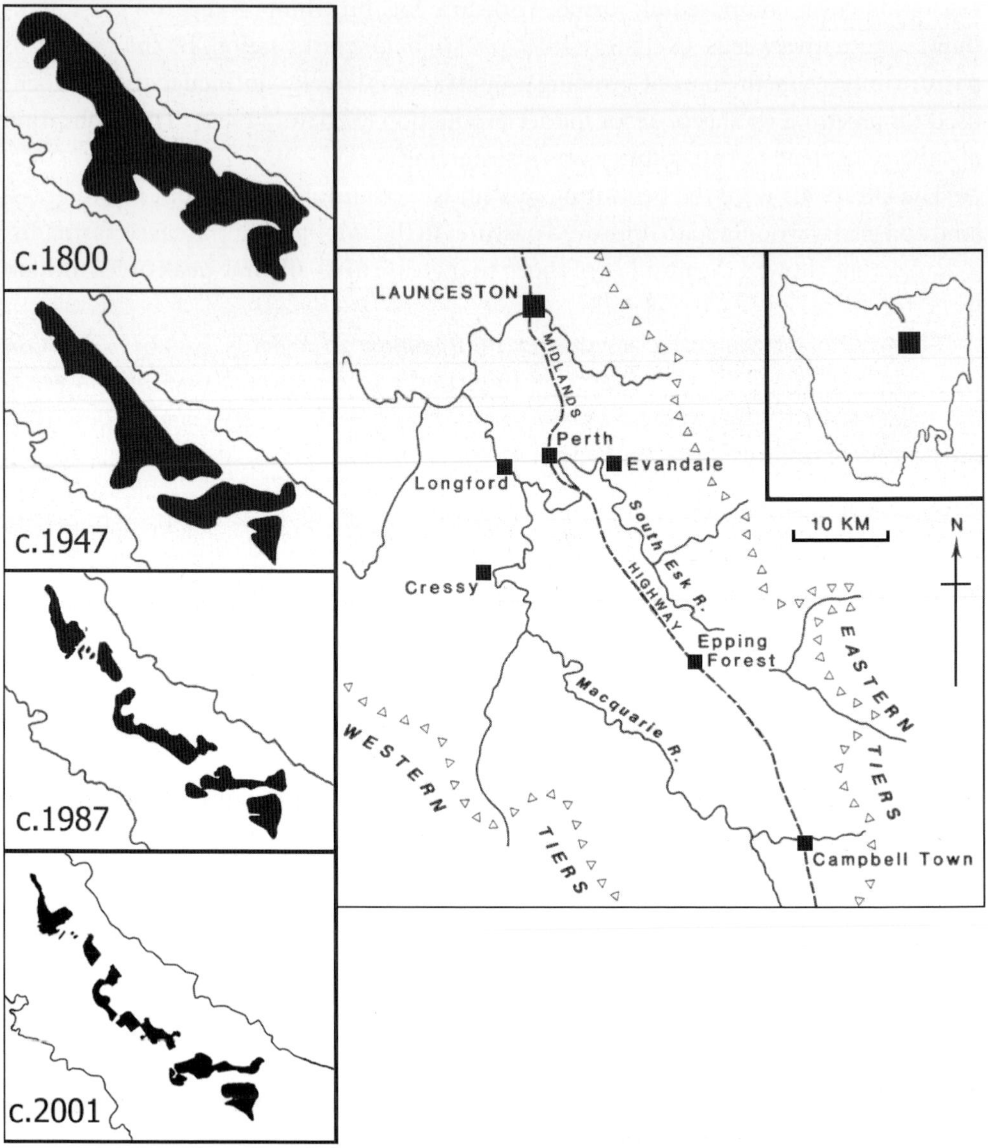

Figure 1.20. The disappearance of Epping Forest (modified from Fensham 1991).

Fensham 1991). At the time, there were also substantial tax incentives for clearing. These were removed in 1983, slowing the rate of land clearance for improved pasture throughout the wool-growing areas of Tasmania (Kirkpatrick and Dickinson 1982; Kirkpatrick 1991).

Land clearance and pasture improvement had many ecological consequences. They rendered some previously common vegetation types and plant and animal species rare or threatened. For example, the type of black peppermint (*Eucalyptus amygdalina*) forest found on the gravel-rich sands of the Midlands is recognised by the Forest

Figure 1.21. Remnants of black peppermint forest on Tertiary sands and gravels, around Conara Junction, looking east into the Fingal Valley, 2005 (Jamie Kirkpatrick).

Practices Board as a type that should no longer be cleared because of its rarity and depletion (see Figure 1.21). There is also a concentration of rare or threatened plant species in the native vegetation remnants in the driest parts of the state. On the other hand, the improved pastures have proven highly attractive to some native species, especially the brushtail possum, white cockatoos, corbie grubs and cockchafers (*Oncopera* and *Aphodius* spp.), all of which appear to have dramatically increased.

Markets

Wool was the mainstay of the Tasmanian economy in the second half of the nineteenth century. The 1843 crash, and consequent low demand and prices, was followed by a steady rise in prices until the late 1860s. Prices then fell to the mid 1890s, to rise once more until the early 1920s. The following fall became precipitate in the Great Depression that followed the crash of 1929. Nevertheless, in 1933/1934, 12 per cent of the value of all production in Tasmania was attributable to wool (Lowndes and Maze 1937). Wars in cold climates have always been good for wool sales and prices, with the Korean War inducing a price peak for the twentieth century. When wool prices are adjusted by the CPI to 1989 values, the late twentieth century becomes a tale of dramatic decline, from the peak of 1784 c/kg in 1950/1951 to generally below 400 c/kg from 1990/1991 onwards (as shown in Figure 1.22). In 1972/1974 and 1987/1989 prices increased, but only to fall to much lower levels in the following years (see Figure 1.22).

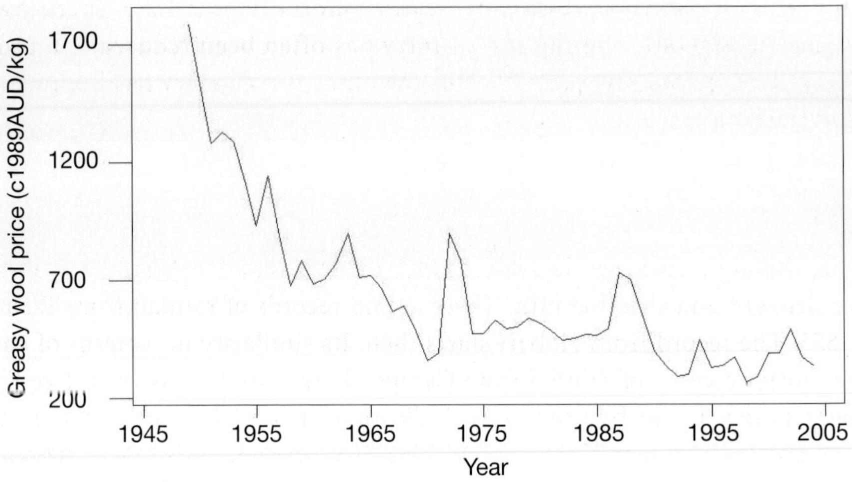

Figure 1.22. Consumer Price Index adjusted wool prices since 1950.

Given this trend, and the dramatic increase in the cost of labour that occurred during the twentieth century, it is no wonder that properties that once supported villages of more than 50 workers are now run by one or two people, their dogs and the occasional contractor. It is also not surprising that, in the period of lowest prices from 1990 onwards, sheep numbers in Tasmania declined from their peak of 5.3 million to 3.3 million (see Figure 1.5).

The major peaks and troughs of the economic cycle tend to precipitate changes in land use, which have consequent effects on nature. Economic theory suggests that depressions usually lead to a consolidation of properties in the hands of the most efficient. Booms provide capital that can be used to initiate changes in management that might counter the later declining prices. Fences and a thriving stud industry rose from the disaster of the 1840s depression and the following price peak. The depression of the 1890s and the later period of high prices may have helped initiate the subsequent era of pasture improvement. The Great Depression of the 1930s and the early 1950s peak in prices presaged the era of massive land clearance and pasture improvement. More idiosyncratic economic events can also have major effects on land use. For example, the calling in of payments from the many grazier 'names' in Lloyds of London in the 1990s resulted in the sale of many wool-growing properties and subsequent intensification of land use under debt pressure. The period from 1990 onwards has been one of extremely low wool prices, but not a period of general economic depression. The decline in wool prices since 1950 has been more a function of the increasing market success of artificial fibres than a reflection of economic cycles.

One response to historically low prices has been that Tasmanian wool-growing properties with any land suitable for irrigation have diversified into cash crops, particularly potatoes and poppies. Paddock trees have been removed to allow the free movement of pivot irrigators, and dams built all over the landscape, often drowning remnant native vegetation in the run country (Figure 1.21). Sheep grazing pressure on the vegetation of the run country has often been reduced. On properties without land suited for cropping, the low input run country has become more critical to the enterprise.

Climate

Tasmania, although to a lesser degree than the northern island of Australia, is a land of alternate drought and sheeting rain. There are no records of rainfall from Tasmania before 1855. The record from Hobart starts then. Its similarity in patterns of change with the shorter records of rainfall from Campbell Town and Launceston (see Figure 1.11) suggests that it can be used as an indication of rainfall history in the sheep-growing districts of Tasmania. The period 1855/1856 was relatively dry. Between 1857 and the mid 1890s there was only an occasional dry year, with a similar pattern between 1917 and 1977. There were long runs of dry years, with just the occasional wet year, between the mid 1890s and the mid 1910s, and between 1978 and the present. Dry years tend to have less late summer–early autumn rain than wet years (Kirkpatrick *et al.* 2000). This makes drought problems worse than would be the case with a seasonally even reduction in rainfall, because plants are most subject to moisture stress during the warmer parts of the year.

Variations in climatic conditions have not had a major effect on the total Tasmanian wool clip, or the number of sheep, being overwhelmed by other factors, largely prices and technological innovation. Warm, dry years undoubtedly cause economic distress for many individual landowners, because they cause a shorter growing season than otherwise where rainfall is normally low. However, where rainfall is normally high and where temperatures are normally low, the growing season can be extended in dry warm years. Thus, losses can often be balanced by gains at the statewide scale.

Droughts have had some major effects on nature in the wool-growing country. Around 1910 trees were reported to be dying in the paddocks of the drier parts of the Midlands. Most of the survivors of this dry period died in the harsher dry period after 1978. Where not removed for pivot irrigators, dead trees now scatter paddocks where live trees stood in the early 1970s. The dry period since the late 1970s seemed to play a major role in the reduction of burning in the run country of Tasmania, with autumn burning being much more risky than in earlier times.

In the same period, many wetlands that, on advice from landowners, were thought to have been permanent by Kirkpatrick and Harwood (1983a,b) proved to be highly ephemeral. Permanence was obviously contingent upon the continuation of the climatic conditions that prevailed between World War I and 1978. Lake Dulverton, which had been seriously considered for a rowing course of international standard in the early 1980s, dried out completely. As a consequence, the Great Crested Grebe (*Podiceps cristatus*) ceased to breed in Tasmania.

Figure 1.23. A Midlands landscape similar to the pre-European conditions. Note natural salt scalds in foreground (Jamie Kirkpatrick).

Conclusions

The Tasmanian wool-growing industry of the early twenty-first century has many attributes that can be traced back to the years of its inception in the first half of the nineteenth century. A large proportion of the families that took up land at that time still own and work wool-growing properties, and live in stone mansions built by convicts. Many of the sheep that are shorn today are descendants of the great rams bred in Tasmanian studs in the late nineteenth century. Nineteenth century ecological mistakes, such as the introduction of gorse and rabbits, are an ongoing source of management expense. Yet, some of the landscapes in the run country differ little in their appearance from the pre-European condition (Figure 1.23).

In the twentieth century increasing efficiency in the production of wool was enabled by a series of technological developments, including mechanisation of farm activities, fertilisation and seeding regimes, effective biological and chemical controls for pest organisms and methods for rapid and cheap conversion of native forest and woodland into improved pastures. These developments have greatly reduced the native vegetation of the wool-growing country (Figure 1.1a–f), and have had major effects on the populations of native animals. Nevertheless, throughout the history of wool growing in Tasmania, native pastures have provided an important resource in terms of both fodder and shelter, and still have a place in wool growing in the early twenty-first century. The next chapter describes the nature and importance of that place.

Chapter 2

Managing the run country for production

J.B. Kirkpatrick, K.L. Bridle and P. Leith

Summary

While wool growing is an important component of the economy, few wool growers make a good return on capital from wool alone, relying more, where they can, on opium poppies, potatoes, prime lambs, wood chips, tourists and cattle. Potential investments in the productivity of the runs equal the somewhat risky potential returns. The runs are regarded as economically valuable for fodder, particularly for sheep producing premium fine micron wool, and shelter. They are managed in a wide variety of ways, partly reflecting the nature of properties, and partly as a matter of choice between alternative systems, based on perceptions of risks and returns. Rotational grazing and set stocking systems have different challenges and benefits, as do burning and not burning. There are different ways and degrees of controlling wild herbivores and weeds. There is no one right solution.

Introduction

It is a challenging task to make a living from the land. Using complex native systems as the base for production increases the level of challenge. Runs are not featureless plains, like much Australian cropping country. Individual properties differ markedly in their sizes and environments. On the runs, the manager of the enterprise must be cognisant of the qualities of their land and the native vegetation it supports.

The Australian Bureau of Agricultural and Resource Economics (ABARE) reported that there were 400 specialist sheep farms, and 308 mixed enterprise sheep farms in Tasmania in 2004. The typical specialist sheep farm sold 24 484 kg of greasy wool, while the typical mixed enterprise sheep farm sold 9655 kg. The mean total capital value per specialist sheep farm was $2 297 000, with a mean total debt of $184 000. The equivalent figures for the mixed enterprise sheep farm were $2 460 000 and $159 000. The mean price per kilogram of greasy wool was approximately $5.00, making the mean gross returns from this commodity per specialist sheep farm approximately $122 500 and the total contribution of wool from specialist sheep farms to the State Gross Domestic Product $49 million dollars, supplemented by $15 million from the mixed enterprise sheep farms. While not providing a substantial return on capital investment by itself, wool growing certainly contributed substantially to the economy of Tasmania.

While most wool is a product of improved pastures, wool growing relies on native pastures to a greater degree than other agricultural activities. These native pastures range from vegetation closely similar to that present in pre-European times to degraded improved pastures in the process of invasion by native species. Vegetation types, as defined by Harris and Kitchener (2005), that are commonly grazed by sheep in Tasmania are: highland *Poa* (tussock grass) grassland (Figure 2.1); lowland *Poa labillardierei* (tussock grass) grassland (Figure 2.2); lowland *Themeda triandra* (kangaroo grass) grassland (Figure 2.3); lowland grassland complex; rockplate grassland (Figure 2.4); lowland grassy sedgeland; highland grassy sedgeland; *Eucalyptus amygdalina* (black peppermint) forest and woodland on dolerite (Figure 2.5); *Eucalyptus amygdalina* forest on mudstone; *Eucalyptus amygdalina* inland forest and woodland on Cainozoic deposits; *Eucalyptus globulus* (Tasmanian blue gum) dry forest and woodland (Figure 2.6); *Eucalyptus gunnii* (cider gum) woodland; *Eucalyptus ovata* (black gum) forest and woodland; *Eucalyptus pauciflora* (cabbage gum) forest and woodland on dolerite; *Eucalyptus pulchella* (white peppermint) forest and woodland (Figure 2.7); *Eucalyptus rodwayi* (swamp peppermint) forest and woodland (Figure 2.1); and *Eucalyptus viminalis* (white gum) grassy forest and woodland (Figure 2.8).

Figure 2.1. Highland *Poa* tussock grassland with *Eucalyptus rodwayi* woodland (Jamie Kirkpatrick).

Figure 2.2. Lowland *Poa* tussock grassland on this side of a deer-proof fence (Jamie Kirkpatrick).

Figure 2.3. *Themeda* tussock grassland (Jamie Kirkpatrick).

Figure 2.4. Rockplate grassland (Jamie Kirkpatrick).

Figure 2.5. Black peppermint forest on dolerite (Jamie Kirkpatrick).

Figure 2.6. *Eucalyptus globulus* woodland (Jamie Kirkpatrick).

Figure 2.7. *Eucalyptus pulchella* woodland (Jamie Kirkpatrick).

Figure 2.8. White gum (*Eucalyptus viminalis*) grassy woodland (Jamie Kirkpatrick).

This chapter documents the ways in which the native vegetation of the runs is used in wool-growing enterprises, the benefits to the enterprise of this use, the management problems that beset those who use the runs, and the ways graziers feel about, think about and solve these management problems.

Runs, native pasture and bush

The definition of 'run country' varies somewhat from farmer to farmer. The following definitions of run country and types of run country sit in the middle range of this variation. Runs are the parts of a property that are largely covered by native species, and which are used to graze or shelter sheep. Runs are usually dominated by native grassy vegetation, and are typically located on ridges, slopes and rocky sites on 'poorer' soils that are unsuitable for improved pasture development. Run country usually consists of large paddocks. Run country may be native pasture without any trees. It may also be 'semi-improved', in that it has had fertiliser added and been aerially sown with clover: 'We have 1500 acres of run country that has been supered and aerially sown with sub-clover.' 'Degraded pasture' in the runs is a pasture that has undergone some improvement in the past but has reverted to a weedy native pasture. Run country also includes 'bush', that is, areas of native vegetation with trees and shrubs.

Sources of information

The major source of information for this chapter was discussions with the managers of 48 wool-growing properties in the period 2003–2005. Henceforth, these managers are usually called graziers, although they are sometimes called wool growers or wool producers for the sake of variety. The discussions were designed to document the knowledge, opinions and feelings of graziers on the ways in which they managed their sheep and native country. They also covered the problems and benefits attached to this management, and observations on native ecosystems and species. The discussions were informal, but gently directed towards the above ends. They usually involved a tour of the property and a cup of tea, during both activities the discussions were either recorded or transcribed. Quotes from these discussions are used widely in several chapters of this book.

Additional information was collected on rainfall, altitude, property size, the proportions of the property consisting of cropland, improved pasture, native pasture and native bush, and the stocking rates on different types of pasture. How the farm is managed may depend on the opportunities available to the grazier. These two sets of information allowed for a comparison of opportunity and management for the 48 properties. Statistical techniques were used to help in this process, and to develop a classification of the stated practices and purposes of management systems.[1] This classification, described below, is a device to simplify discussion, rather than an attempt to develop a set of widely applicable 'farming styles', as reviewed by Vanclay *et al.* (1998, 2006), although those familiar with this literature will probably be able to cross-correlate between classifications.

The properties discussed tended to be larger, and almost certainly had a smaller proportion of cropland and improved pasture than sheep-growing properties in the state as a whole.[2] However, they cover a wide range of size and degree of development, except for those properties that are totally improved.

The microeconomics of run country

> 'To improve the country would cost more money than we'd get in return.'

The most recent analysis of the economics of wool-growing enterprises was undertaken as part of the 8 x 5 Wool for Profit program as a benchmarking exercise. The two financial years in which Bob Reid and Associates (2003) undertook this analysis were both years of low wool prices (see Figure 1.22) and recovery from drought. It is therefore not surprising that the returns on capital for the farms as a whole was generally higher than the return on capital invested in wool production, with means of 4.9 per cent for farm and 4.0 per cent for wool for 2000/2001 and 7.2 per cent and 5.7 per cent in 2001/2002. Nevertheless, many graziers got a better return on their

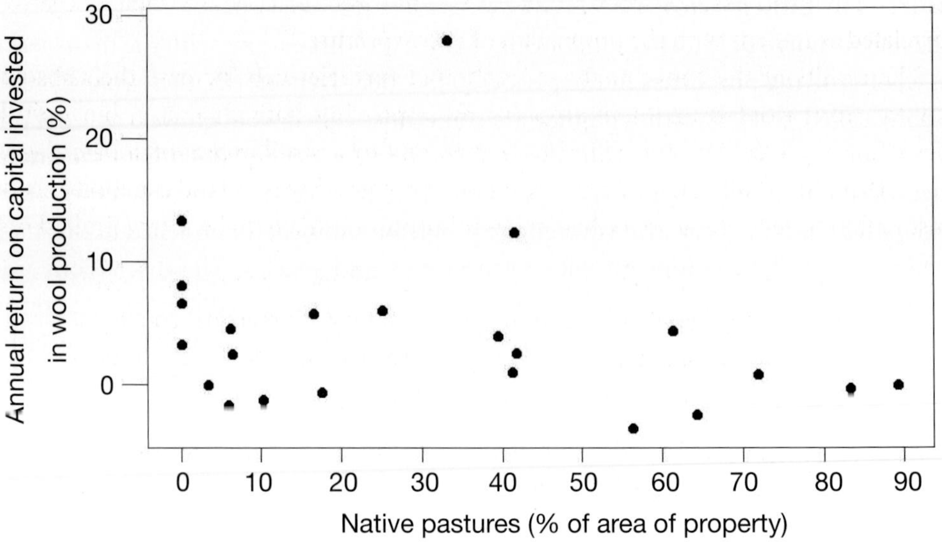

Figure 2.9. The lack of a relationship between annual return on investment in wool production and percentage of the property consisting of native pastures (data from Bob Reid and Associates 2003).

investment in wool production than on their whole farm investment. In 2000/2001 30 per cent of farmers gained more than 5 per cent return on investment in wool production, and 54 per cent of farmers did the same in 2001/2002. The focus of the 8 × 5 study was on the enterprise rather than the role of run country in the enterprise. However, data comparing the proportions of properties under different types of land use between the 20 per cent of the properties which had the highest return on investments in wool production and the rest could be taken to indicate that the most successful properties had lower mean percentages of native pastures than the rest (2000/2001, 24.5 per cent compared with 17.1 per cent; 2001/2002, 21.3 per cent compared with 16.8 per cent). However, the difference between these figures is not statistically significant.[2] The relationship between return on investment in wool production and the percentage of the property consisting of native pasture was also poor[3] (see Figure 2.9). Thus, the proportion of the property consisting of native pastures had no statistical effect on the return on capital invested in wool production. This implies that, at the time of the study, investment in improvement of native pasture was generally less attractive than investment in blue chip shares and about equal to investment in a term deposit. Real returns on wool have not increased since 2001/2002, and the general tendency is downwards (see Figure 1.22), suggesting that these conclusions still pertain for pastures used for wool production. However, increasing prices for prime lamb may work in the other direction.

The data from Bob Reid and Associates (2003) (in Figure 2.9 above) show that a smaller proportion of native pastures in a sheep-growing property results in significantly higher: dse per grazed ha; greasy wool production (kg/ha); clean wool production (kg/ha); wool returns ($/ha); wool gross margins ($/ha); and, wool gross margins

($/dse). However, costs of wool production ($/ha) were equally as strongly negatively correlated as returns with the proportion of native pasture.[4]

The results of the above analyses imply that a grazier can increase their absolute returns from wool from their property by improving native pasture, but only by increasing their capital and running cost inputs to a similar proportionate degree. Given that almost all native country used for sheep grazing is in land capability classes unsuited to development (see Chapter 7), this option is likely to be a very limited one:

> '... everything that can be turned into good pasture has been improved. The natives [vegetation] are in the stony country, it's marginal country. It doesn't pay to clear any more land any more, and I couldn't afford to do it now. It's better to put more into the improved country.'

> 'Ploughing costs $350/ha to make new pasture, and takes two to three years to get established – is it worth it?'

Some graziers only work their run country because of debt pressures:

> 'Financial pressure, if we didn't have this, we could cut the stocking rate way down low and make life easier for yourself and shut up the native country.'

It seems likely that differences in return on capital investment in wool growing might largely relate to attributes of the property for which farm-level data were not provided by Bob Reid and Associates (2003), such as property area and the quality of the land. Wool micron, staple strength and management efficiency also undoubtedly play a role in the variation in return on capital.

There is no doubt that wool growing has been an economically difficult activity in Australia in the first decade of the twenty-first century, because of low wool prices (see Chapter 1, Figure 1.22) and drought (Leith 2006). In Australia, in the period 2001/2002 to 2003/2004, the average rate of return, excluding capital appreciation, for the top 25 per cent of specialist sheep farms was 2.7 per cent; for the remaining 75 per cent it was –1.6 per cent (ABARE 2006).

Economic uses of run country

Run country is used by wool producers for a variety of economic purposes, the primary one being feed for stock. The importance of runs in the workings of the enterprise may exceed their direct contribution to the bottom line:

> 'It's different in Tasmania, we have highly variable production. We have very productive land but only if it rains. We can't manage for the bad years else we'd never make money, and we can't manage for the good years else we'd flog the place so we have to manage somewhere in between. The run country allows us to do this. The run country is an important component of the whole farm. We stock wethers in the bush in autumn/winter and it's spelled in spring/summer. Stocking the runs in

autumn/winter allows the 'lower' (improved) country to be rested as there isn't a lot of growth on it. Bringing the wethers down for shearing and then putting them on the improved country takes advantage of the excess growth during spring and gives the bush a rest at a time which is ideal for that country. Stock eating the excess growth maximises the productive use of the pasture. Stock put on weight before they go back into the bush. When they come down for shearing they have lost weight. I'd say we can run an extra 3000 sheep because we have the bush. Most of the flock will go towards paying costs, but it's the extra sheep that make the profit. So it's hard to put a value on the native country, because you'd have to cost in the relieving of the improved country as well as the number of stock the native country runs.'

Others work it because it is the only thing they can conceive to do with it:

'It's not hugely profitable – it gives the country some value but there's not much else you can do with it.'

The runs have a widespread good reputation among graziers for producing wool of a fine and consistent micron without major inputs:

'(We use runs) to produce superfine wool from wethers. If you put them on improved pasture, it increases the micron by 1–2.'

'I'm not super fine but we're all looking for ways of trying to even out the highs and lows in nutrition which will lead to an improvement in staple strength. Native vegetation generally doesn't have the ups and downs. This is important for the wethers (which are a dead loss, I wouldn't have them if I had the country for ewes). Grassy plants offer even nutrition, maybe not the shrubby ones though.'

'If we put them back out fat and they stay out there, they come in quite well. If they go out in poor condition, they'll struggle and there's no value in it. We grow our best wool on that country. We grow volume down here and quality out there.'

Wool from runs is regarded as largely chemical-free, and runs are valued for cleanliness in terms of disease and contamination of wool by dust:

'Soils in the bush are ironstone/shallow soils. It grows very clean, very white superfine wool.'

One grazier commented on problems with a break of thickness resulting from the flush of growth in spring on improved pastures. One wool producer was worried about high levels of mortality on the runs, and another placed coats on his sheep when they were in runs to keep them clean. Several graziers noted problems with mustering in bush runs, which differ from native pasture runs in their high densities of trees and shrubs.

Part or all of both native pasture and bush in the run country is used by many graziers as a drought reserve:

> '*Poa* and sagg tussocks are my hayshed, my standing drought reserve.'

The major native grasses are much less responsive to variation in water availability than the sown and adventitious exotic grasses that dominate improved pastures. In particular, kangaroo grass (*Themeda triandra*) maintains good growth during summer, when most exotics brown off. This means that improved pastures and runs can be highly complementary, with more sheep being able to be supported without feeding on properties with a pasture mix than on those with only one type of pasture (Mokany *et al.* 2006).

Probably because the use of runs for feed was felt not to need comment, fewer graziers mentioned it than the use of the runs for shelter, particularly for lambs and off-shears. Saggs (*Lomandra longifolia*) and tussock grasses (*Poa* spp.) were particularly valued for shelter, although both, when large enough to provide shelter, provide little feed for sheep, and therefore are often regarded as undesirable components of all but the run paddocks used for these purposes.

A small number of graziers mentioned commercial use of their run country for timber (5) and firewood production (1). Although the bush in the run country is dominated by trees, these tend for the most part to be extremely slow-growing, so only a subset of properties in higher rainfall country are suited to long-term silviculture. Wood chipping income is still used by some to develop country that would otherwise have stayed native:

> 'We will keep developing some country that was woodchipped and is overgrown with wattles. We'll rechip it and leave the good young trees and all the poor area. We'll super and seed and improve the country.'

> 'It's hard to justify doing anything but the chips will pay for the improvement.'

Firewood collection is undoubtedly more widespread on properties than is indicated in the discussions. On one property, close to Hobart, the owner had a steady income stream from royalties on firewood gained from paddocks full of trees that had been ringbarked in the past. A few graziers mentioned gaining an income stream from charges to deer hunters for the use of their properties. Some gain an income from tourism, particularly 'colonial' or 'farm' accommodation.

Managing the run country for profit

The enterprise classification[1]

There is always more than one solution to any one combination of opportunities and problems and Tasmanian wool-producing properties are highly variable in both. It is therefore not surprising that there is a wide variety of ways that managers graze sheep in the run country of Tasmania. The graziers fall into four major groups, named by a mixture of characteristic attributes of their systems and the environments of their properties (see Table 2.1).

The first group, the 'improved country set stockers', had high percentage frequency values for set stocking, fertilisation of native pastures and the use of sagg/tussock country for off-shears and lambing (see Table 2.1). Their properties were concentrated in the Midlands, had low rainfall, a high proportion of improved pasture and a low proportion of bush (see Table 2.2).

The second group, the 'improved country rotators', had high percentage frequencies for rotational grazing, cropping, irrigation and using the bush for shelter (refer to Table 2.1). Some fertilised native country, but most did not state that they did so. This group was geographically widespread. Their properties tended to occur in low altitude and low rainfall areas and had a high proportion of improved pasture (see Table 2.2).

Graziers in the third group, the 'high and moist country graziers', largely used a combination of rotation and set stocking, rested their runs in spring, had a growing focus on prime lamb production, had recent changes in management and mostly did not say that they fertilised their runs (see Table 2.1). Their properties occurred in areas of high altitude and rainfall, and they had a high proportion of bush and a low proportion of improved pasture (refer to Table 2.2).

Those in the fourth group, the 'warm country graziers', were almost equally divided between set stocking and rotational grazing. They did not say that they fertilised native pastures, widely utilised drenching, had Saxon Merinos, used their sagg/tussock country for shelter during lambing and commonly rested their runs in spring (refer to Table 2.1). Their properties were concentrated at low altitude in the South East and Northern Midlands and had a high proportion of bush and a low proportion of improved pastures (see Table 2.2).

Grazing regimes

Introduction

Part of the art of managing runs for profitable wool production is to use a grazing regime that promotes the type of pasture most suited to enterprise and environment. When you ask wool producers how they manage grazing on the 'run' country, they will often tell you: 'it depends', a phenomenon well illustrated by the results of an experiment:

> 'We had an experiment on the effectiveness of wallaby fencing. He (the researcher) set a stocking rate for the sheep, but the sheep did better in the block with the wallabies in it. The important result was that we should have had a higher stocking rate, then the grasses wouldn't have become so rank but we needed to run stock so that we wouldn't have to destock during the experiment. And then we had good seasons. It just highlights the difficulty of being prescriptive with grazing.'

Producers know what they want from the bush runs, what they want from their sheep and for the property and they need to balance all their goals within prevailing

Table 2.1. Percentage frequency of comments related to the wool-growing enterprise for four styles of enterprise

Responses that occur in at least 20 per cent of the enterprises in at least one group are shown. The highest percentage frequency for each comment, and values over 50, are shown in bold.

	1	2	3	4
Number of properties	(12)	(16)	(9)	(20)
Fertilise/super native country	**75**	37	33	0
Set stock	**75**	12	0	**55**
Sagg/tussock country shelters off-shears	**42**	12	22	0
Little change in management over the years	**33**	6	0	0
Lick/urea blocks no good/don't use	**33**	12	11	0
Native pasture as shelter	**33**	6	0	9
Horehound an issue	**33**	19	0	9
Runs cleaner/less parasite problems/less inputs	**33**	25	22	18
Feed ewes prior to lambing	**25**	0	0	0
Don't jet ewes	**25**	0	0	0
Eco-marketing requires perseverance	**25**	6	0	0
Worm problem	**25**	12	22	0
Use bush as drought reserve	**25**	19	0	18
Mentioned lice	**33**	12	**33**	9
Rotational grazing	0	**87**	44	45
Wallabies a problem	**58**	**75**	**56**	**55**
Cropping	25	**62**	22	9
Possums a problem	33	**62**	44	36
Irrigating	8	**50**	11	9
Use bush for shelter	25	**44**	11	0
Deer a problem	16	**44**	33	18
Thistles an issue	33	**37**	02	7
Cats a problem	0	**31**	0	0
Use bush for feed	17	**31**	11	27
Seeding native with introduced species	8	**25**	0	0
Gorse problem	**75**	**50**	**78**	36
Rest runs in spring to allow growth/seed	8	6	**67**	**64**
Combination of set stock and rotation	25	12	**56**	18
Recent changes to management	0	6	**56**	9
Growing focus on prime lamb	0	12	**44**	0
Use native pasture in winter	8	12	**33**	0
Use all/part native pasture as drought reserve	25	19	**33**	18
Ragwort problem	8	0	**33**	0
Rest runs in summer to allow growth	0	19	**33**	9
Avoid dusty conditions/maintain ground cover	0	6	**33**	0
Have whole farm plan	0	12	**33**	18
Corriedale x Merino	0	0	**33**	0
Hunting amenity ($ for hunting rights)	8	0	**22**	0

	1	2	3	4
Use 1080	17	6	**22**	0
Always clean heavy equipment before work	8	0	**22**	0
Paterson's curse an issue	0	19	**22**	0
Focus on superfine from native pasture	17	0	**22**	18
Fluke problem	17	0	**22**	18
Wattles seem to affect health	0	6	**22**	0
Keep grazing charts	0	0	**22**	0
Lease/agist into extra land at times of drought	0	0	**22**	0
Strategy = pasture focused	0	0	**22**	0
Shifting focus of graziers to irrigation	0	0	**22**	0
Reducing stock numbers didn't reduce cash flow	0	0	**22**	0
Graze improved summer	0	0	**22**	0
Rest runs for variable periods	8	0	**22**	18
Use native pasture in summer	0	0	**22**	0
Burn then spell	0	0	**22**	0
Rabbits a problem	0	0	**22**	0
Already use environmental accreditation in markets	0	0	**22**	0
Current farm viability affects perspective	0	0	**22**	0
Commercial forestry in bush	0	6	**22**	18
Fire won't carry in permit season	0	0	**22**	9
Have decreased stocking rate on native pasture	0	0	**22**	18
Drench sheep	33	19	22	**64**
Sagg/tussock country shelter for lambing	42	12	0	**45**
Saxon Merinos	25	0	0	**45**
Fine wool from runs more consistent	0	6	0	**45**
Kangaroos a problem	8	25	11	**36**
Better grass cover in runs keep wool cleaner	8	6	0	**36**
Urea/lick/supplements used in dry times	0	12	11	**36**
Try to spell some runs for reserve	8	0	11	**36**
Use native pasture all year	6	6	0	**27**
Young wethers not on runs/on best runs	8	19	0	**27**
Usually burn in spring	8	6	0	**27**
Rabbits not a problem anymore	8	0	0	**27**

1 = 'improved country set stockers', 2 = 'improved country rotators', 3 = 'high and moist country graziers', 4 = 'warm country graziers'.

climatic conditions. A dry year can have a severe impact on grazing management for years into the future, not just the following year (Leith 2006). To add to this complexity and climatic contingence, there are fluctuating numbers of wild herbivores that impact on which area is grazed, and when and how it is grazed. Experience has taught

Table 2.2. Environmental, land use and locational characteristics of properties in enterprise groups with ANOVA results for means

	1	2	3	4
Mean altitude (m)	286ab	237a	453b	182a**
Mean annual rainfall (mm)	558a	562a	678b	627ab*
Mean property area (ha)	5252	2645	5319	3665 NS
Mean % bush	15.6a	26.2ab	49.8b	53.8b**
Mean % native pasture	27.5	11.8	28.1	19.4 NS
Mean % semi-improved	13.8	13.6	0.0	8.0 NS
Mean % improved pasture	42.3b	44.0b	14.3a	16.1a**
Mean % crop	0.0	2.7	5.6	2.1 NS
Central Plateau (number)	1	1	2	0
Northern Midlands (no.)	5	9	5	5
Southern Midlands (no.)	6	3	2	1
South East (no.)	0	3	0	5

1 = 'improved country set stockers', 2 = 'improved country rotators', 3 = 'high and moist country graziers', 4 = 'warm country graziers'. ** = $P < 0.01$, * = $P < 0.05$, NS = not significant. If any letter to the right of a figure is the same in a row, then the two values are not significantly different at $P < 0.05$.

wool producers about the carrying capacity of particular runs and when is a good time to move stock to a different run:

> 'When to move sheep. Well, you look at their condition, how much fodder there is, how short the pasture is.'

> 'Blue grass, tussock grass, we put the cattle in if it's too tall. If it's too short, you know you have too many sheep.'

Influence of environment on pasture management

Most producers have an opinion on what is the best grazing system for the country and for their own purposes. This opinion is formed from experience, knowledge and capacity (labour and infrastructure – fencing and watering points) to undertake the various options. Opinions vary for very good reasons. One striking example, related to environment, is the variety of views on the value of kangaroo grass. Kangaroo grass is killed back to ground level by severe frost, but persists year round in a green state in many low altitude and coastal areas. The frost-killed foliage has a much lower nutritional content than the green foliage: '*Themeda* is good until it gets frosted then the stock won't touch it.' For this reason, kangaroo grass is valued more near the coast than inland, accounting for variation in comments on the species from 'kangaroo grass is an indicator of sward health' to 'abundance of kangaroo grass indicates insufficient grazing pressure'. (Appendix 1 shows the wide-ranging views on grazing management.) Those who valued kangaroo grass had properties with a mean altitude of

182 m, whereas those who did not make this statement had properties with a mean altitude of 313 m. The relative abundances of kangaroo grass and wallaby grass in pastures can be influenced by season of spelling, kangaroo grass appreciating spring spelling and wallaby grass appreciating winter spelling (Garden *et al.* 2000). Kangaroo grass will also become less abundant if overall grazing pressure is increased (Garden *et al.* 2000; Kirkpatrick, Bridle *et al.*, 2005; Kirkpatrick, Gilfedder *et al.*, 2005).

Set stocking

The major elements in a grazing regime are timing and stock numbers. One extreme in timing is set stocking (see Figure 2.10) in which a low number of sheep, usually wethers, are left on the runs throughout the year, except for a period in which they are shorn, usually spring, when they are often held on improved pastures:

> 'The ewes are almost always on the improved country. I have a stable number of wethers that are basically set stocked on the run country. I've got 3000 acres divided into three runs, about 600 acres, about 1000 acres, and about 1500 acres. They would be gathered in October for shearing, and depending on that season I would hold them in the paddock for one to three months to give them a break, to pick up body weight and to let

Figure 2.10. Sheep in set stocked run, with saggs (*Lomandra longifolia*) in foreground (Jamie Kirkpatrick).

> the runs take off a bit. It is always a delight to see the kangaroo grass come up and harden off a bit, before the sheep are put back up there. But, if the paddock is needed for other sheep, the bush wethers could be back up their within a couple of weeks. So, it's nine to 11 months of the year that they are on there.'

Advantages of set stocking are the low input cost of maintaining the stock and low infrastructure requirements.

> 'We don't do rotational or cell grazing. You need different fences and watering points. Our stock are watered by the creek, permanent water.'

One producer states that set stocking suits his needs and those of the sheep.

> 'They get to know their own run, wethers, and when you take them away from it they quite often want to get back there too. So we've sort of stuck to the old-fashioned system of set stocking with a summer break wherever possible to let the grasses seed out. And it works pretty well, it's simple and effective and we know, depending on the season, how much stock to put out there each year.'

> 'Wethers get to know where to go for a feed. They get confused if they're shifted. It's just a matter of the right stocking rate.'

Graziers who said that they set stocked tended to also say that they had Saxon Merinos; had wethers in their runs all the year; had little recent change in management; believed that runs were cleaner, with fewer disease problems and fewer inputs than improved pasture; thought that urea blocks were no good; and, had not fenced their rivers and streams.

Rotational systems

The other extreme to set stocking lies in rotational systems, some of the more intense of which, like cell grazing, involve the placement of large mobs of sheep in small areas for a short time period, with a long rest period before sheep are reintroduced (see Figure 2.11). Most regimes are intermediate, involving periods of spelling, either on an annual basis in a particular season or seasons, or when it is judged by the grazier that the pasture in a paddock needs a rest. Resting is not necessarily absolute, some graziers reducing stocking rates in response to a perceived deterioration in native pastures, rather than removing stock altogether (see Appendix 1).

Graziers who said that they used a rotational system tended to also say that they varied their stocking rates according to the type of country:

> 'Our grazing management consists of sheep eating the north-facing slopes out in spring, then we remove sheep late November, then get feed/seed set. We move the sheep to the back hill, and then to the major part of run which is 600 acres of varied wooded/open country and the sheep are

Figure 2.11. Mob of sheep in rotationally grazed run (Kerry Bridle).

> there for most of the time. The 600 acres get about 10 weeks rest from shearing and while the sheep are in the other two paddocks. The north-facing slopes get about two weeks grazing and the back hill gets the rest. The rotation may occur twice a year. I value the vegetation for holding the soil in place. You don't get problems.'

> 'The native cell helped us with our grazing management. The sheep used to hang on (the) front hill, now we can move them and spell the area.'

Those who said that they used a rotational system also tended to say that they had initiated recent changes in management. Many reasons were given for the change to rotational systems:

> 'We don't do set stocking. There are too many negative effects such as sheep tracks, nutrients, stock camps, uneven grazing pressure, rank vegetation and bare ground. Our degraded pastures are made up of natives and annual weeds. They were probably never sown, but were set stocked for the last 100 years.'

> 'In the days of burning and set stocking, rabbits prepared the beds for wattle establishment.'

> 'In the shady areas there is fog grass. The sheep will live on one side. If you run larger mobs and break the paddocks in half, this would improve the quality (of grazing) and cover in preferred areas and fight off weeds.'

> 'I used to set stock Merino wethers in the bush, now I rotationally graze them and have increased the stocking rate even though they're on there for a shorter period. The land is out of production for half of the year, but production from the other half is compensating for it. Strategic grazing is grazing when you really need it.'

> 'We initially set stocked – but when the season got tight we had no flexibility. Now we rotate and generally try to leave one to two runs empty for flexibility. We learnt. We thought you could put more on than you could. For good wool the balance has to be right – the grass ... doesn't come back as quickly (if it's too heavily stocked).'

The recent increased use of rotational grazing in runs is theoretically grounded in the proposition that this system results in greater utilisation of the less palatable species and sections of runs, and thereby does not select against the more palatable species. This should result in increased pasture production from rotation of runs compared to set stocking. This increased production did not eventuate in the only experimental test of different grazing regimes in Tasmania, with spring spelling producing significantly greater net pasture growth than both rotational grazing year-round, and rotational grazing in autumn and winter, on a kangaroo grass/wallaby grass pasture near Nile in the Midlands (Friend *et al.* 1999). The growth from the two rotational regimes was statistically indistinguishable from continuous grazing (Friend *et al.* 1999). It needs to be noted, however, that this pasture had been previously managed using spring spelling, so that its particular species composition may have adjusted to this regime. A poor soil seed reserve of palatable species (Gilfedder and Kirkpatrick 1993a,b; Friend *et al.* 1997) makes for a slow response to changes.

The rotational system does solve the problem of heavily eroded north-facing slopes within paddocks, eliminates sheep camp effects, and is highly effective in mustering in difficult country, as the sheep tend to be very keen to move to the next paddock:

> 'Big mobs of sheep need good water sources, especially where the wethers are – out the back outside the catchment area. Waterholes are all full this time of year, but we also have a few troughs in the run country. You can't set and forget, and you need to make sure there is enough feed. Moving them is not a problem. The sheep are ready to move when the paddock is eaten out.'

However, wild animals can be always one paddock ahead of the sheep:

> 'The bush run gets more native grazing than sheep grazing because roos are one step ahead of the sheep. It's a good season this year because the roo numbers are fewer due to recent droughts, but I'm not getting the relief I thought I would.'

The institution of a rotational system usually involves substantial investment in new fencing and the provision of new watering points as well as a greater investment of time:

> 'It's (rotational grazing) more time consuming than set stocking at lower rates.'

Some producers have tried rotational grazing but found it to be of no advantage to them.

> 'We started a few years ago shifting wethers around on the run country, and we sort of got caught in the end of not having anywhere to go, because normally there's two or three equal size areas that carry roughly an age group, so after doing one of these grazing courses we decided to group them together and move them every few weeks, but come the middle of winter we weren't really gaining anything, there was nothing growing, so we sort of got caught in a bit of a hole, and it was confusing for the sheep more than anything. If they would have been left alone set stocked doing their own thing they probably would have been fine, but because they were grouped up and being shifted every few weeks at a time when there wasn't much feed, it was a bit tough on them.'

Cell grazing

The philosophy and practices of cell grazing have been taken up by several graziers:

> 'Our change of thinking was encouraged by the RCS (Resource Consulting Services) regular meetings with guest speakers. We heard about the cell grazing focus with a long-term approach which encouraged you to work on triple bottom line, not just economics. The whole concept originated in Zimbabwe and South Africa, based on the behaviour of migrating animals.'

> 'For cell grazing you need a 90-day rest rotation in hard times, in good times you can speed that up. For stock health, you have to move them regularly, no more than three days grazing in a paddock. You need 30 paddocks minimum. We only had 12 paddocks initially and that was no good. To do well, we had to have 30 paddocks. One of my cells is 40 paddocks. The first day the sheep eat the ice-cream, the second day they eat the brussel sprouts. If you have them in there over 7–10 days then that's a disaster because of the micro-organisms in the gut. And the root systems improve and get away over a 90-day rest but 120 days is preferable.'

Having said this, the same producer then comments:

> 'Cell grazing is not really a proposition on native country because you can't get the water to them (stock).'

The cell grazing approach gives producers more control over their stock. Set stockers tend to use runs as a drought reserve, whereas cell grazers are more likely to destock if a drought is imminent.

> 'We have grazing charts and we can offload sheep and be flexible in our stocking rates.'

> 'Cell grazing helps ration tucker. If it doesn't do anything for the pasture it helps grow more grass. I have no drought reserve. My policy is to look at the grazing chart and look at the feed on hand and the rainfall and stock or destock when necessary.'

This grazier talks about cell grazing, but is referring to improved pastures. However, he gives a good overview of the principles behind cell grazing:

> 'The sheep have one week on and 11 weeks off according to grass growth. You need to maintain cover and assess the sheep. I use the following – winter ration, spring utilisation, summer clean up. You need to adjust stocking rates for the kind of season that you're having. We use grazing charts to predict feed availability, using a 12-month rolling average rainfall. I try to adapt cell grazing principles to suit me. I did the RCS course and I like the degree of control that cell grazing gives you. Ideally I want to have smaller paddocks and larger mobs and decrease inputs by using stock to control weeds. I want to build up organic matter in the lighter soils through my grazing management. I run one-third ewes, one-third wethers and the balance works well. I can be more flexible with my grazing management and can move the wethers off the farm. I use grazing charts to collect information on what paddocks will run and if I change my management, then the charts will show whether it's of benefit.'

One grazier talked in detail about his experiences in cell grazing native country:

> 'The runs are grazed for approximately one week three times a year. So you need 12–20 runs on the property for rotational grazing to work. There were two changes that caused me to change my grazing system. The first change was after we did the initial fencing in the '80s, which put the sheep into totally different areas, but in the '90s when we went to the rotational grazing we then improved our control of the grazing. So initially, we were just increasing the size of the farm, if you like. But now it's all about control, how long you graze it for.'

Another grower gave an in-depth account of his change to cell grazing on native country:

> 'I changed to a mixed age group in the mid '80s and used runs for a few years and then adopted cell grazing … . The problem with cell grazing is

that you get a lot of vegetation growth, but it's not useful to stock due to palatability problems.'

Other graziers detailed the benefits they saw from adopting cell grazing:

> 'I stopped burning with cell grazing, to keep the nutrients in the system.'
>
> 'The benefits I see are that cell grazing gives the eucalypts a chance to regenerate because the stock aren't eating them off when they're small. They have a chance to grow when the run isn't utilised, so they grow beyond the palatable stage to sheep.'
>
> 'Cell grazing gives you more vegetative growth and less bare ground, and uses less labour. Ten years ago, this site had less than 30 per cent ground cover, now it's kangaroo grass, which I use as an indicator of grazing pressure.'
>
> 'We've got our ground cover back, so that's all we'(ve got) now, a lot of it isn't really desirable species and things, but hopefully, and I think, well, it is happening, we're getting some succession change, and hopefully that will keep going. But I was more than happy to initially get the ground cover back, and then we can go on from there. Because we've gone from here out to there (moving stock into areas they didn't graze before). And that's another thing, there was no kangaroo grass out there, and now it's coming back.'

By changing grazing systems, producers find that they must also allow for and manage other changes in their run country. There can be problems with cell grazing:

> 'One of the things that's happening now, and we're going to have to watch it, is because the eucalypt canopy is opening up, because of dieback, we're now getting lots of wattles, and especially with the cell grazing.'

Cell grazing also requires a higher degree of concentration on management than other systems:

> 'Just one thing on the fencing, with our set stocking system it required a lot less management. It was an elastic system. In the good seasons, the sheep were (on the better areas), and in the bad seasons they went out (into the rougher areas). But they absolutely pound the good areas before they go out into the rougher country. But once you increase your fencing, fencing off north and south slopes, and you increase grazing on your south slopes, so you're actually increasing your stocking rate, but when the bad seasons come, your management has to be more elastic. Whereas before, the environment was elastic. You can't just set and forget. You have to be more careful.'

One producer:

> '… tried cell grazing years ago. There's a lot of work involved in rugged country. I had sheep on both systems (cell and set stocked). The cell grazed sheep did better with body weights but there was no difference in the wool cut. They came off runs in better condition but gave the fences a punishing. They had a bigger body weight to maintain which meant they were hungrier sheep. The paddocks worked okay in an unfertilised system but the growth rates are slow. Paddocks need rests of up to 120 days. You can't consider native woodland management as a cell. You need to manage grazing according to the weather.'

Another who had tried it said:

> 'For cell grazing you've got to be on the ball. You've got to manage it all the time. You've got to do it (move sheep) else you're in trouble. You couldn't do it in this country (in runs) anyway. We had a bad run. We didn't manage it properly and got a lot of tender wool. Set stocking is more flexible. Some farmers had trouble with it (cell grazing), one of the problems is that the sheep get into a routine, after about two days they've eaten the best of it out so then they'll start hanging at the gate waiting to go into the next one.'

Others, who had never practised cell grazing, perceived problems with the system:

> 'I think cell grazing will be a disaster in wet years because of footrot (spreading it around the property).'

> 'I'm not anti-cell grazing, the trouble is the country is heavily timbered and I'd be living on a horse (to move sheep). I'd block graze on the open country.'

> 'Cell grazing, I wouldn't touch it, it's not the way to go for superfine wool.'

> 'Cell grazing, if we had big open areas then it might work but young sheep hate being in big mobs.'

> 'I think there are advantages in NOT cell grazing. It's very expensive to set up, you need a lot of water, and I've heard that the production rates are not quite as good. But if you have the money to try these things out … . I'm not thinking of going that way myself. You can do cell grazing without setting up that great rigmarole of a watering system and fences.'

The above quotes make it apparent that cell grazing is a controversial technique, but one that many graziers feel is a productive innovation. It may be less suited to runs than to improved pasture.

Mixed systems

Many graziers, especially those in the 'high and moist country' enterprise group (Table 2.1), said that they varied their grazing systems between set stocking and rotation according to circumstances:

> 'It's certainly not cell, it's a combination of set stocking and rotational grazing… . Well, we have done all three, and to a greater extent we still amalgamate them. It depends on what the grass levels and the seasons are like. If you take a drought, whether you do cell grazing or set stocking, you still have a drought. In lush conditions, cell grazing really comes into its own, if you're looking at gross returns over a five-year period… . (In) mid-winter … I would be set stocking for lambing, and for the rest of the time a combination of all three.'

> 'It's not relevant to think of the difference in different grazing systems on native country, everyone does a mix of all of them. You don't always have the choice. Some areas are unsuited to any particular system so you use a mix. It's better to look at the principles of grazing such as: avoid overgrazing, there's less damage if you understock and the bush works well as a drought reserve.'

One grazier described a mixture of approaches related to the fertility of different parts of his run:

> 'On the native pastures out to the west where we run the wethers it is set stocked, but on the native pasture hill country that's had more regular fertiliser, that's rotationally grazed. It depends on what's there before they go in, so it's not overgrazed by the time they go out. It's mainly used for young sheep on this type of country. It's spelled more during winter when the ewes come down onto the more arable ground for lambing. And once we've finished with the lambs around Christmastime, they go back into a rotation around the hills again.'

Spelling

Spring was the most frequently mentioned time for the spelling of runs:

> 'Only these last few years when we've had dry years that we haven't been able to do it but normally all bush wethers stop in home to eat rubbishy grass until Christmastime – gives bush a rest. It's had since the middle of October until now before sheep go back out.'

Those who said that they spelled their runs in spring also tended to have said that they had a whole farm/property plan; had instituted recent changes in management; ran Saxon Merinos; believed that runs could have a sales advantage in that wool from this source was totally or largely chemical-free; valued *Poa*; and, had possum problems.

Those with high percentages of improved pasture tended not to say that they rested runs in spring, whereas those with high percentages of bush country did tend to say that they used spring spelling.

Summer was the next most commonly mentioned season for spelling runs after spring, followed by winter (see Appendix 1). No-one mentioned spelling runs in autumn. For other graziers, resting was associated with particular events, such as after shearing, during lush years or in dry years. One grazier said that he kept stock off his runs for several years after a drought. Five graziers noted that they rested runs for variable periods. The spelling process was often quite complex:

> 'I use *Poa* tussocks as an indicator species, as soon as that gets within 6" we move the stock off. Sometimes a run can be spelled for 12 months to give the grasses a chance to recover. Native country is fairly forgiving; as long as it's given enough time to recover after it's been hit hard. Sometimes we hit it hard – it usually takes three dry seasons for us to do this – and there's usually a break before then and we can spell it for 12 months. Some places where the tussocks have gone, with a couple of years of careful grazing they come back.'

Those who said that they used all or part of their runs as a reserve for drought, flood or other contingencies tended to also say that they had kangaroo grass or tussock grass. Some graziers bought in feed to enable the resting of runs in drought years.

Many graziers mentioned the times of the year that they had sheep in the runs (see Appendix 1). Not surprisingly, all of the year was the most commonly mentioned strategy, followed closely by winter. Those who said that they used runs in winter also tended to say that they had problems with thistles, perhaps a consequence of ground disturbance at this time. A few graziers mentioned using their runs in summer. One of these put on his sheep in large numbers for a short period. One grazier said that he stocked his runs in autumn.

Type of stock

There was also variation in the type of stock used on runs in different circumstances (see Appendix 1). Many graziers, particularly those in the 'warm country' group, said that they kept young wethers off their runs. These graziers tended to also say that they had *Themeda*, drenched, had a covenant and had possum and wallaby problems. While wethers as a whole were kept on many runs all year, especially by set stockers, young wethers were preferentially kept off low altitude runs (mean of 124 m versus 307 m). Three graziers mentioned that they had both young and old wethers on their better runs. One grazier placed his young ewes on native cells. Another confined his hoggets to the runs. Those, largely 'warm country graziers', who said that they used runs with saggs and tussocks for lambing also tended to say that they: had wallaby grass and *Themeda*; thought that the runs provided better grass cover; used to burn but now saw it as too risky; had problems with weeds coming in with feed; drenched; and, were involved in the North-Facing Slopes Program. Those, largely 'improved country set stockers', who said that they used sagg/tussock country for shelter for off-shears also tended to say that they focused on producing superfine wool from native pastures; had an increasing focus on prime lambs; fertilised their native pastures; valued the cleanliness and lack of diseases on runs; thought environmental accreditation would be valuable; thought that the lack of chemicals in wool from the runs might be valuable in marketing; were regenerating part of their bush; were involved in the North-Facing Slopes Program; and, had worm problems. Three graziers said that they put their off-shears in the bush. Several graziers observed that they took wethers off the runs and onto improved pasture to fatten before being sold.

Saxon Merinos were the most mentioned breed of sheep used in the runs (see Appendix 1). Those, largely the 'warm country graziers', who mentioned this breed also tended to say that they set stocked; rested runs in spring; used the bush as a drought reserve; did not fence streams; regarded the risk of burning as too great; and, had problems with weeds in feed. Corriedale Merino crosses were mentioned by several of the 'high and moist country' group of graziers. Border Cross sheep were as frequently mentioned but not concentrated in an enterprise group, with White Suffolk, Polwarth Cross and Cormo sheep mentioned by one grazier each. The various breeds have their virtues in different situations:

> 'Border ewes we've got 2000–3000; when pasture is dry they will eat horehound; Merinos (Saxons) won't eat it. Corriedales would starve on a bush run – that's the difference in breeds; some need to be well-fed. Saxons were out of fashion in the '70s but are now back in because of their finer wool. You can't buy Saxons. Saxons like (to eat) a mix of everything.'

> 'They (Saxons) are natural browsers, they live on bark and scenery.'

Several graziers mentioned their mixing of sheep and cattle in the bush runs (see Appendix 1). Cattle were regarded by some as valuable in the reduction of coarse tussocks and for opening up bush, these processes sometimes being facilitated by the use of urea blocks.

Stocking rates

Stocking rates in runs in the 48 properties were mostly in the range 0.5–1.5 dse/ha. Apart from their magnitude, stocking rates were little mentioned. Six graziers said that they wanted to increase the numbers of stock on their runs, while four others had recently reduced the numbers of stock on their runs. Two of these latter graziers observed that a reduction in stock numbers had not reduced their cash flow. One grazier said that he stocked heavily to keep the micron low, while another observed that low stocking rates are good for sheep health. Some graziers were cautious with stocking rates:

> '... better to be understocked than over then you've got no real worries. My father ... used to bring 5 to 6 to 8 hundred sheep out here (1000 acre

run), put them out here for two to three months a year. They'd eat the run right out then move them on; eat the next run out and then the next one. Bush country takes so long to recover; once you eat them out it's a long long time to recover; it's not like a grassy paddock. Especially in the wintertime; there's no way in the world that it's going to recover.'

Others maximised their stocking rates:

'An economically conservative stocking rate is not as good as higher rates and coping with drought.'

A grazier commented on difficulties in planning stocking, related to climatic variability in Tasmania:

'I farmed in WA. The principle's the same but the seasons are predictable in some areas; you nearly do things on the same date each year. Eastern Australia is so variable from season to season, feed is so variable with rainfall … can't stock to a predetermined program.'

Fire management

The use of fire as a tool to maintain runs in a productive state appears to have been almost ubiquitous on runs until the 1970s. Fire was often used to open up woody areas for grazing. Wattles and prickly box were seen as a particular problem:

'You need a fire to thin the wattles down. A hot fire will kill wattles. It will cook the bark and there will be no resprouts. Only small wattles can be killed with a cool fire.'

Prickly box could also be reduced:

'The lack of fire has led to prickly box getting away in a run. Now I can't use the run, it has too much wood.'

Another grazier commented:

'I won't burn now. I used to light a match and leave the fire to burn. It would remove the box and other unpalatable vegetation.'

Keeping the bush open was good for two reasons,

'more regular fires encourages grasses',

and

'the thicker the bush, the more severe the fire. If you burn more regularly, it kills the young trees.'

This grazier also noted that the reduction in fires has led to 'the bush thickening up'. Many graziers observed that burning was valuable in turning rank saggs and tussocks into green pick, producing grass cover and reducing woody plants. One grazier suggested that burning helped clear the runs of worms. Another suggested that it might help prevent insect attack on trees. On the other hand, some graziers worried about charcoal in wool:

> 'Wool buyers didn't mind charcoal in wool at one time but not now.'

Effects on nesting birds, and fires causing tree death were also noted (see Table 2.3 below).

Table 2.3. Percentage frequency of comments related to fire management

Don't burn	35
Used to burn but stopped	19
Burn saggs and graze after fire	17
Risk of burning too great	12
Hazard reduction burns mainly/remove undergrowth/firebreak	12
Usually burn in spring	10
Usually burn in autumn	8
Burn saggs	8
Fire promotes good feed	6
Usually burn end of winter	6
Need to burn but difficult to manage (time)	6
Fire won't carry in permit season	6
Charcoal in wool a problem with burning (market)	6
Small burns prevent big ones	4
Reserves are fire hazard	4
Concerned about birds/nests with burning	4
Controlling burn difficult	4
Burn/intends to burn native pasture to control tussock	4
Burn then spell	4
Lack of grazing in bush increases fire risk	2
Try not to damage trees with burns	2
Hot fires threaten biodiversity	2
Fire breaks cause erosion/stopped making them	2
Concern about tree death from fires	2
Small patch burns only	2
Regular burning encourages grass	2
Graze heavily to reduce fuel load near road (improved)	2
Fire necessary to control heathy plants in grassland	2
Burning helps clean runs of worms	2
Used to do bigger burns/now only small burns	2
Fire might reduce insect attack on trees	2
Cool fires prevent eucalypt regeneration	2
Keep stock out after burning	2

In the late 1970s, the combination of a prolonged reduction in late summer/early autumn rains, and greater restrictions on the use of fire, made many graziers reconsider its use as a management tool. Twenty-one of the 48 graziers said that they did not use fire at the time of the discussions. Nine of these 21 stated that they used to burn, but had stopped. One said:

> 'If you flog the land hard enough you don't need to burn.'

Six of these nine graziers thought that the risk of burning was too great (see Table 2.3). The lack of fires in recent times may have led one grazier to opine that:

> 'It's too hard to control fires. We don't have enough resources.'

Many others commented on the legal ramifications of an out-of-control fire:

> 'A neighbour lit a fire in the hills over there. We didn't try and sue him. We just put the fire out. You don't sue your neighbours, it's just not done.'

A grazier in the Northern Midlands noted that:

> 'The Fire Service took over the rural part and started promoting not burning. In this area everyone expects their neighbours' fire to go over the boundary some time.'

One family was reluctant to burn as it was too time consuming to get a fire to carry outside the permit season:

> 'If you try to burn, it's too hard (they're too busy). There's a fine line between wasting time and getting too much (a hot fire). We need to find the time, it needs doing.'

Not all graziers have been sufficiently discouraged to stop burning. Eighteen of the 48 still burn, although six of them stated that they used it for hazard reduction (see Table 2.3), implying that they did not burn for production purposes. Some of the graziers who continued to use fire showed evidence of deep thought on the matter:

> '... burning is a very important part of grazing management. As well as reducing fuel, I firmly believe that we are following the Aboriginal fire regime to a point. I believe their fires were cooler fires, and that if we have less frequent fires, then we are threatening our biodiversity.'

Comments from many graziers indicated that skills in fire management were highly valued:

> 'My father was non-stop burning every time he went out but he never caused a bushfire. He prevented plenty. There's a fine line between a sensible fire and a stupid fire.'

Those burning to promote grasses and reduce competitive woody plants adopted a variety of regimes. Five said that they burned in spring, four said that they burned in autumn, and three said that they burned at the end of winter (Table 2.3). One grazier remarked:

> 'I used to burn behind mustering – you could tell who was mustering by the fires.'

Some graziers burned in two seasons. One said:

> 'I go out in the spring and autumn. Autumn is more difficult. If you go out in the spring as it is drying out, you can try and try and try and the fire just won't run, or if (it does) it is a very cool fire, very patchy and light. Two days or a week later – you have a very narrow opportunity to burn – the fire will be moderate or hot, and then either the fire restrictions are on, or it's no longer safe because the soil is dried out too much, and the vegetation will burn too hot. So, yeah, there are areas that I haven't burnt for all those reasons, and the eucalypt regrowth stands up to all that.'

Another grazier commented:

> 'I used to carry out medium intensity burns … any time it was safe – late spring or autumn.'

Several graziers said that they burned in small patches, or burned dense stands of particular plants, particularly ferns, saggs and tussocks. Some graziers said that they spelled their runs after burning. The period between fires seemed to be adjusted according to the need, the fuel load and available time to do the burning. One grazier commented:

> 'I used to burn in rotations of five years. This was regulated by what you could get to burn. But with the decrease in rainfall in the '90s, it would be a longer interval now because the vegetation isn't growing.'

Another said:

> 'So I try and burn my country – the open run a little bit more – but between a three and a five-year frequency. That doesn't mean it happens, it is bureaucratically harder and harder all the time, to burn, because of the pressures we are put under, and the threat if it gets away, and adjoining landowners, and forestry are not burning, and I rely on natural boundaries and my own knowledge and the weather to control fires largely, and often, you will keep trying, and fires won't run, and then they will run, and you only have a few days or a week before it is too hot, and you can't burn anymore. So there is a very narrow window of opportunity. So, often various areas don't get burned for much longer periods, generally 15 years.'

Six graziers mentioned that they only used fire for hazard reduction, the reality being that run country can become more flammable in the absence of fire, especially in good seasons when stock and wild animals cannot keep up with the growth. Stock were used to create low fuel conditions around assets:

> 'Stock are kept on the cleared land around the house and sheds in November/December as a firebreak and to spell the bush.'

Bush blocks outside the kitchen window were also subjected to grazing to reduce the risk of fire over the summer months:

> 'We put 650–700 wethers on the native block. It cuts the fire risk down. You've got a bomb on your back door if you don't graze it. That's the biggest trouble with 'the reserve' (covenanted block). Cattle chew the tussocks down a bit. They decrease the fire risk.'

Changes in grazing regimes have reduced the need to burn on some properties:

> 'I used to burn all the time, but I haven't since the introduction of cell grazing about 10 years ago.'

Some graziers feel obliged to protect their properties from fires lit by others. One said:

> 'My property is dissected by the highway. So my principal goal at the moment is eating out the paddocks on the highway as hard as I can in case of fire … I'm building myself a safety unit.'

Another commented:

> 'Instead of little fires, you now get blokes who light fires in permit time and the rest of us go out to fight it.'

However, sometimes unplanned fires are seen to be beneficial:

> 'Ten years ago a "kind" fire went through the 1400 acres. It was a cool fire lit by the neighbour, though he denies it. Another neighbour said that if he didn't light it, it's the only one he didn't light!'

Another family commented on a fire that escaped on their own property:

> 'There was really good feed in the bush after that stubble fire. It was the best thing that could have happened.'

Increasing productivity through inputs

Eighteen of the 48 graziers said that they used fertilisers on their run country. Only one grazier stated that they did not use fertiliser in runs, although four stated that they did not use fertilisers in their bush runs, and three said that they had used fertiliser, but had stopped (see Table 2.4). Two graziers said that they applied superphosphate from the air. Those graziers who said that they applied fertiliser tended also to have

Table 2.4. Percentage frequency of comments on increasing productivity through inputs

Fertilise/super native country	37
Problem of weeds coming in with bought feed mentioned	19
Urea/lick/supplements used in dry time/or to encourage browsing of less palatable plants	15
Lick/urea blocks no good/don't use	15
Increasing/want to increase stocking rate on native	12
Use urea/lick blocks for runs	10
Super advantages introduced species/weeds	10
Seeding native with introduced species	10
Don't fertilise in treed country/bush	8
Used to fertilise native country/no longer	6
Feed ewes prior to lambing	6
Aerial spray has killed trees	6
Urea/lick/supplements used in winter	4
Super encourages native trees	4
Regret improving some native/run areas	4
Plough native	4
Fertiliser encourages grubs	4
Do soil testing for P application	4
Cost of super limits usage	4
Aerial super	4
Use trace element supplements	2
Urea blocks aid mustering	2
Supplements an expensive fertiliser	2
Super not good for natives	2
Super encourages native grasses	2
Sowed tussock species into improved	2
Removing saggs	2
Involved in 8 x 5	2
Hard to know how often to fertilise runs	2
Fertiliser use must be carefully targeted	2
Feed sheep as part of normal management	2
Feed sheep during dry periods	2
Ewes sometimes have mineral deficiencies	2
Don't fertilise/super native country	2

said that they seed the run country with exotics; urea blocks/licks were no good; they used saggs/tussocks for shelter for off-shears; they were participants in the North-Facing Slopes Program; and, wombats were found in their runs.

The positive association with wombats suggests that fertilisers may be applied more in the sandier, less fertile runs that favour their burrowing activity, rather than the more fertile runs on more clayey soils:

'The wombats seem to like sandy country, rather than stony country, because they dig holes.'

The heavier stocking allowed by fertiliser application may have exacerbated erosion problems on north-facing slopes where paddocks were topographically variable. Fertiliser application can stimulate the growth of *Poa* tussocks (Groves *et al.* 1973), thereby not affecting the value of runs for shelter:

'Super affected the native species. Some stayed like wallaby grass, along with the clover and rye grass. The *Poa* tussocks thickened up and needed to be burnt out, they shaded the other species too much. A lot of the improved country we drove through still has a very good cover of *Poa* and sagg. They've done well with super.'

Fertilising, sometimes followed by seeding of the run country with introduced plants, is a stratagem designed to increase the production of highly nutritious species, particularly clovers:

'We have 3500 acres of semi-improved runs. The areas that are cleared of trees are top dressed (super and subclover) and are predominantly native grasses. Top dressing does increase the productivity. The semi-improved open grassy runs run one ewe to an acre.'

'If you increase fertiliser, then you increase the stocking rate. You need to do that to pay for the fertiliser. But you do need some now and then, to put back what you take out from harvesting the wool.'

It was suggested that fertilising improved nutrient cycling:

'... one of the obvious things is that the sheep shit doesn't break down as well, and to me it seems that the nitrogen cycle isn't working properly, it just stays there in little dry pellets for ever and a day whereas on fertilised country where you got the nitrogen cycle working properly the manure seems to break down.'

In some ways, putting fertiliser on runs can be a bit of a gamble:

'At the moment, the money for the super is coming from capital. If we get an autumn break, I will get the money back, if we don't then I will have blown the lot.'

'It's a bit tricky. Clover is a bit notorious for leaping away in the autumn and that can create a break in the wool. But we tend not to get a prolific autumn break in Tasmania. But that's probably the danger with having clover.'

'If you have a good season like we did last year, you're probably less inclined to put more fertiliser out, and if you know a drought is imminent, I'm more

> likely to put more fertiliser on. So you judge it round that way, but you don't always get it right.'
>
> 'When I developed my country, I didn't see any increase in carrying capacity for many years. The benefits went out the window. Then we had a good year and got an increase and then the increase in carrying capacity holds. But I get the same result from cell grazing and, anyway, half the fertiliser goes to the roos.'

Mokany *et al.* (2006) report experimental work on native pastures at three Tasmanian sites that resulted in substantial increases in pasture production and dse/ha. They calculated that the expenditure in seed and fertiliser resulted in an increased return per ha by the third year. However, they, and several of the 48 graziers, pointed out several problems, related to production, with the fertilisation strategy. Fertilisation usually increases the biomass of weeds and exotic annual grasses in the pasture:

> 'Some of our back runs were supered during the '70s, but it's hard to make the sandstone country productive. The grasses don't persist due to very dry summers and wildlife. The paddock has gone to browntop and fog grass and doesn't produce much.'
>
> 'Native pastures are not worth fertilising. Ours have gone backwards with fertiliser, more weeds, annuals, etc. It's better to double the use of fertiliser on your improved.'

Both the weeds and the clover are more responsive to variations in the availability of moisture than the native grasses, creating potential problems with breaks in wool micron, and reducing drought and summer reserves. This is exacerbated by the fact that the deep-rooted native perennial C4 kangaroo grass, that provides fodder during summer, is strongly disadvantaged by fertilisation (Groves *et al.* 1973; Friend, Thompson *et al.* 1997).

Fertilised native pastures require much more careful management than unfertilised native pastures. Clovers can out-compete the native grasses, requiring strategic stocking to avoid the eventuality of sporadic clover cover interspersed with weedy annuals and the occasional unproductive perennial exotic:

> 'You can use fertiliser and clover and maintain *Themeda* but you have to be careful with the grazing. The loss of native vegetation may be a result of overgrazing rather than the impact of the fertiliser itself.'

Two graziers observed that fertilisation encouraged grubs in their native pastures. One grazier observed that fertiliser was good for native trees and grasses, while another observed that it was bad for natives. It undoubtedly depends on how much, when and into what type of pasture. This is probably why one grazier remarked that it is hard to know when to fertilise runs and another pointed out the need to carefully target its application (see Table 2.4).

Some just cannot afford the upfront investment:

> 'My father poured the super on and the country responded. We would like to put more on to run more stock, but we're still catching up after the drought. We had four years of drought from 1996–2000. The cost of super prevents us from using it.'

Others see the fertilisation of run country as a poor investment:

> 'You're better to add fertiliser to improved country rather than native. You don't know how much control of total grazing pressure you have in native country.'

> 'I don't like chemicals, overall. There is no doubt that our country is low in phosphorus, for the improved species that we are trying to grow. I know that the native species are fine, but we are trying to improve production, and so yes, particularly when I needed more production, when I had a growing family, and I had financial needs. Then I put fertiliser on, and I saw results. And I think it is only really economical to put fertiliser on the better soils. The more marginal in production terms that the country is, the less response you will get. I've stopped putting it on second quality paddocks, if you like. Even there, you are often just encouraging extra wallabies with increased fertility. So heavy stocking goes with heavier fertilising. So my philosophy is that I get lighter and lighter and lighter stocking, and I think a lot of the fertiliser money is not well spent.'

> 'If you fertilise you get more sheep with clover, but not the same quality of wool.'

Another grazier commented on the wildlife problems he saw as attached to fertilisation:

> 'In 1958 we never had a problem with roos (wallabies). I'd lived here for five years and never seen a roo in the bush. The young bloke (worker) is a shooter and he shot 200 last month. Super makes a cafeteria right outside their door.'

Those who set stocked and fertilised tended to say urea blocks were no good:

> 'Have used urea blocks to nil effect.'

> 'During drought we used protein blocks. Can't see that they made a difference.'

One grazier, who supered his native pasture, remarked that urea blocks were an expensive form of fertiliser. Nevertheless, an equal number of graziers, largely 'warm country graziers', used urea blocks and/or licks as abhorred them:

> 'If you need to force them to eat during drought, use urea blocks.'

Urea blocks may be of more use in rotationally grazed systems than set stocking systems, especially when used as a substitute for burning, and in dry times. Salt licks were used by some graziers:

> 'We use salt licks in bush; put them out crutching time.'

Mechanical inputs were used by some graziers. Two said that they ploughed up their native pastures. However, another two graziers said that they regretted improving some of their native pastures. Another was in the process of removing saggs. Yet another was mechanically removing prickly box (*Bursaria spinosa*). Silver wattle and other trees are part of a major mechanical removal program on one property:

> 'Show you what we're trying to develop without disturbing the countryside. All this country down here was all scrub and wattles. We came in here with a wheel machine, not a bulldozer, didn't disturb the soil at all, we just cleaned it all up. The only problem we had was after two to three years was the wattles coming. We were thinking about coming out here and spraying them but we thought no, we'll try not to. So all we've been doing now is slashing it, and we're actually getting on top of it. We've left it all – most of the trees were just about to fall over or dying and everything else was just wattle scrub (trees pushed into a pile and left in the bush) and that's what we plan to do a lot more of in the bush. Let the sunlight into it a bit more, and the native pasture. That's pretty good that pasture – it's been locked up for about six months. … 300 acres cleared area native pasture comes through, more light, you're rid of all dead trees. We use a wheel machine not a bulldozer. It increases the stocking rate, more sun, more pasture. We're able to run 20–30 per cent more sheep. We do about 100 acres per year.'

Disease management

The disease issues most frequently mentioned by the 48 graziers were lice, worms and fluke (see Table 2.5). Footrot, toxoplasmosis and fly strike problems were mentioned by only one or two graziers each. The only one of the three most commonly mentioned diseases of sheep that can have a high mortality rate is fluke (see Chapter 1), which can be avoided by denying sheep access to the habitat of its intermediate host, a snail that occurs in wet, muddy places. It can also be prevented by chemical treatment:

> 'In wet winters there are fluke snail in bush. Stagnant water holes breed them up. Sheep go and drink out of the holes. Only takes two to three

Table 2.5. Percentage frequency of comments related to diseases of sheep and their management

Drench sheep	33
Lice issue	21
Worm problem	15
Wool from runs almost/completely chemical-free	12
Fluke problem	12
Lice problem with natural marketing	8
Wattles seem to affect health	6
Hobby farmers cause sheep health risk	6
Neighbours increase health risk	6
Don't jet ewes	6
Use chemical residue tests	4
Chemical-free is a good marketing tool	4
Mulesing saves chemical usage	4
Monitor wethers for fluke	4
Footrot an issue	4
Fly strike sometimes a problem	4
Drench some years	4
Drench on faecal egg count	4
Dip/backline sheep	4
Cell grazing risk of footrot	4
Wool from high altitude less chemicals	2
Vaccinate for fluke	2
Toxoplasmosis problem	2
No fluke problem	2
Manage fly/lice at shearing as needed	2
Lice not an issue	2
Jet weaners	2
Jet ewes	2
Japanese market wants chemical-free wool	2
Fly strike not a problem	2
Drier seasons break worm cycle	2
Don't introduce sheep (except rams)	2
Don't drench wethers	2
Burning helps clean runs of worms	2

weeks to kill a sheep. You always fluke drench in winter whether you think they need it or not. They can look perfectly good and two weeks later there are dead sheep everywhere.'

Lice and worms can debilitate sheep, thereby reducing wool production, or, in the case of face lice, prevent sales. Because lice are ectoparasites without any intermediate host, it is possible to eliminate them from a flock, if all sheep are thoroughly treated, and if no sheep with lice break in from neighbouring properties to socialise. Worms are almost impossible to totally eliminate from a flock, as their eggs lie quiescent on pastures, awaiting ingestion. They also build up resistance to the chemicals used in drenching, making it important not to over-drench. The worm cysts do not survive dry conditions. They are also densest where sheep have been densest. Thus, one grazier said:

> 'Our grazing management for worms in winter is to put the young stock on the hills.'

However, as in most things to do with sheep management, there was an alternative point of view:

> 'Worm infestations are related to health of animal, nutrition not (stocking) density.'

Problems with worms and lice were preferentially mentioned by the 'improved country set stockers' and the 'high and moist country graziers', while fluke problems were equally as frequently mentioned by the 'improved country set stockers', the 'high and moist country graziers' and the 'warm country graziers', but not at all by the 'improved country rotators', very few of whom mentioned any disease problems.

Both worm and lice problems are generally thought to be more likely at high stocking rates, resulting in the frequent comment that runs are cleaner and healthier than other pasture. The mention of a worm problem tended to be associated with the mention of thistle and horehound problems, both these weeds being indicators of heavy grazing pressure. The mention of a lice issue was also associated with the mention of a horehound problem. Three graziers suggested that wattles were good for health problems:

> 'Wattles fixed scouring sheep.'

> 'Native country sheep/cattle don't get wormy. There may be a natural effect of wattles or their systems might be working better.'

> 'Sheep eat wattles during spring/early summer but ignore them the rest of year. Old days drench made from wattle bark.'

Another three noted the health problems associated with hobby farmers, a comment on the dangers of disease transmission from property to property, supported by another three graziers who said that neighbours can increase the health risk:

> '... only about two or three times in 27 years has there been lice, from the neighbour. You have to be careful with lice because you're damaging the wool and the sheep. So I would backline them.'

> 'Had lice 10 years ago from Richmond sheep; (Richmond), the lice capital – (full of) hobby farmers.'

One grazier remarked that they did not introduce any stock from outside, except for rams, undoubtedly a prophylactic measure.

While footrot was little mentioned, one grazier thought it to be a widespread problem:

> '... half the sheep in Tasmania have footrot. There's none on here. I did have it 20 years ago, but the climate is drier now and there are areas where you can keep away from it.'

A large proportion of graziers, especially 'warm country graziers', said that they drenched, possibly one reason why they did not mention the diseases thereby prevented. One grazier said that they did not drench wethers, while three said that they avoided drenching ewes. Two graziers said that they drenched only occasionally, while another two mentioned that they drenched on the basis of the results of faecal egg counts. Apart from drenching, active interventions to prevent disease that were mentioned were jetting and backlining for lice, mulesing for fly strike, vaccination for fluke and burning of runs for worms. One grazier praised the health impacts of the biodynamic system they used:

> 'Since biodynamics, I've not had to jet sheep with chemicals. Fly strike affects 1:400, this year I've had 2:2000. Last year nothing, year before nothing ... . We'll see what happens next year with super added.'

Many graziers were conscious of the increasing marketing importance of reducing or eliminating chemical residues in wool, one emphasising its importance in the Japanese market and two others seeing a chemical-free state as a good marketing tool. Many graziers noted that wool from runs was largely or totally chemical-free, although four graziers noted that lice were a problem with natural marketing. One grazier said:

> '... no reason for putting anything (meaning chemicals) on a sheep if they don't get lice.'

Two graziers mentioned using chemical residue tests, a strong indication of serious concern. Two graziers defended mulesing as a way to avoid using chemicals, the main short-term alternative.

Wild animal management

The major animal management problem noted by graziers related to wallabies, closely followed by possums, deer and kangaroos (see Table 2.6). All of these animals, as well as the less mentioned wombats, rabbits and hares, eat pasture that could otherwise be eaten by sheep, so have a direct impact on the economics of the run country:

> 'If you were grazing normal pasture – you'd have a level that you'd take it to and not go beyond that. It's amazing what a little bit of cover will do, lifting the wind off the top of the ground. We have the sorts of (wildlife)

Table 2.6. Percentage frequency of comments on wildlife problems and their management

Wallabies a problem	62
Possums a problem	46
Deer a problem	29
Decrease in devil population	25
Kangaroos a problem	21
Increase in possums	12
Bush is habitat for ferals and natives (they increase grazing pressure on pastures)	12
Wombats present	10
Use 1080	10
Cats a problem	10
Increase in cats	8
Rabbits not a problem anymore	8
Increase in wombats	8
Value ecosystem services – control of pasture pests	6
Increase in deer	6
Hunting amenity ($ for hunting rights from groups)	6
Fertiliser encourages grubs	4
Rabbits sometimes a problem	4
Rabbits a problem	4
Increase in wallabies	4
Hunters pull out weeds in bush	4
Don't use 1080	4
Decrease in possums	4
Cockatoos a problem	4
Wood ducks a problem	2
Wombats potential erosion problem	2
Wombats impact on fences	2
Wombats a problem	2
Without control game would take all feed in short time	2
Wildlife restricts ability to use bush	2
Wallabies take fresh pick after burn	2
Shooters only interested in stags	2
Shoot ducks	2
Public access problems with dogs	2
No game management plan	2
Need to develop market for possum pelts/fur	2
Increase in roos	2
Increased game management is a cost of bush	2
Hares a problem	2
Grass grubs a problem	2
Grasshoppers a problem in drought	2
Electric fences useful for excluding native animals	2
Domestic dogs a problem	2

Dogs not a problem	2
Devils an issue	2
Deer not a problem	2
Decrease in roo population after drought	2
Corellas a problem	2
Cats have destroyed birdlife	2
Birds as biological control of grubs	2

numbers where it's eaten into the ground – we're not able to manage all grazing.'

'In my experience, the grazing property has some improvements even if it is largely native, so improved country that is adjacent to habitat is likely to be dramatically increased in terms of wallaby numbers. When seasons are good there is enough for everybody, wallabies, sheep and landholders, but when seasons are tough the wallabies come in more, and when one attempts to close off a paddock to get a bit of production, that is the one the wallabies go to.'

They can also damage infrastructure, particularly fences:

'Wallabies are harder to control. They do damage to fences.'

A higher proportion of the 'improved country rotator' enterprise group mentioned problems with wallabies, possums and deer than those in the other three enterprise groups (see Table 2.1), perhaps because of the long periods in which there are no sheep in run paddocks. The presence of sheep has been observed by some graziers to deter wild herbivores:

'If a paddock is heavily, or is moderately stocked by sheep, wallabies tend to go somewhere else.'

'Wallabies and other wildlife graze where there are no sheep. If you rotationally graze in the bush, it could be that the wallabies are one paddock ahead of the sheep.'

However:

'In a small paddock you could put sheep in and the roos will disappear but in a bigger paddock, I've seen roos in one corner and sheep in another.'

Another reason for the high level of the perception of wild animal grazing problems by 'improved country rotators' may be the tendency of members of this enterprise group to be croppers and irrigators (see Table 2.1). If given the opportunity, most wild animals would prefer to feed on a well-watered and fertilised crop than on the harsh fodder in the run country. Possums are notorious for their mass attacks on some crops, and are much more difficult to keep out than wallabies, deer and kangaroo:

> 'I had a cherry orchard fenced off. It was electrified and irrigated. I went away for a month and the possums had killed every tree, thousands of them.'

Deer also attack crops:

> 'Possums and deer eat new trees. We tried planting seedlings in shelter-belts a few years ago, but they were all eaten by possums. Deer rub their antlers on new trees and they get into crops such as the poppies.'

Those who made specific mention of bush being a habitat for feral animals that ate their native pasture tended also to say that they used the bush as shelter for stock, and/or as a drought reserve, and had a covenant:

> 'Wallabies come out of the covenanted area. We shoot 1000 a year, and it doesn't seem to dampen the population.'

Some covenants restrict grazing by stock. Occasionally stocked bush probably does not receive the same level of management inputs as more continuously stocked bush, potentially allowing a build-up of wild animal populations:

> 'The reserve is a harbour for roos and possums. Possums aren't as bad as they were 10 years ago. In one month a trapper got 3500 skins, when they were worth something.'

> 'Huge numbers of forester kangaroos have come in the last 10 years, several mobs of 50–60 on the farm. The large areas of bush on the neighbouring farm are a wildlife haven.'

Wallaby and possum problems very strongly tended to be mentioned by the same graziers, as, to a weaker degree, did wallaby problems and an observation of increased possum numbers. Those who noted that they had a wallaby problem also strongly tended to be those who did not say that they had ceased burning the bush. One grazier observed that wallabies take the green pick after fire. There are some ecological data that show that wallabies utilise native shrubs for fodder (Dickinson and Kirkpatrick 1986), raising the possibility that they might be a valuable adjunct to sheep in run country where sheep grazing alone might result in shrub dominance. The effectiveness of wallaby grazing in suppressing shrubs is likely to be greatest immediately after fire.

The mention of problems with deer tended to be associated with the larger properties (mean of 6556 ha compared with 2992 ha for those who did not mention a deer problem). Those who mentioned problems with deer also tended to say that tree dieback was an issue; they hazard reduction burned; and, they had problems with fluke. Problems with fluke were also associated with kangaroo problems. Most of these relationships seem more likely to be a product of the largely coincident environmental situations of deer, kangaroo and fluke than causal. The mention of kangaroo problems was concentrated in the 'warm country' enterprise group, again, probably reflecting environmental coincidence, as kangaroos are confined to the more open, low altitude run country and the 'warm country graziers' are more prevalent there.

Deer are protected by law, unlike any other wild exotic mammal, because they are valued by hunters. There are no data on the impacts they have on native vegetation, and they present an expensive management problem, as their access to improved pastures and crops has to be prevented. One grazier has constructed an expensive game-proof fence between his run country and his front country with reasonable success (see Figure 2.2). Others said that they have been discouraged from the best use of some of their land by their inability to eliminate the species. Some graziers, mostly in the 'high and moist country' enterprise group, make a small part of their income from extracting fees from those who wish to hunt deer on their properties, but most regard deer as vermin:

> 'One deer is greater than one sheep; same country grazed. Why are they treated other than like a rabbit? It is time and headache dealing with deer.'

While wombats were not universally regarded as a problem, with the most frequent mention being of their presence, one grazier noted that they presented a potential erosion problem:

> 'Wombats are a potential erosion problem with their burrows. They weren't here 20–30 years ago, now they're in plague proportions!'

Another grazier observed that wombats damaged fences. Four graziers had observed an increase in wombat numbers on their properties.

The rabbit problem of yore is a pale shadow of its former self, with only four graziers mentioning it as an occasional or continuing problem and an equal number noting that they no longer had problems with rabbits. One grazier had a hare problem.

The two main management options for problems with wild herbivores are exclusion or death. Exclusion is usually an uneconomic option within the run country, although one grazier noted success in deterring wild animals from the improved country:

> 'The game fence works. It's four foot high with three hot wires. It won't stop wombats, but it keeps the natives in the bush.'

Death, by shooting or poisoning, is the major adopted option. The wild herbivores that most concern graziers are protected by law, requiring a government permit, or an approved game management plan, before killing can legally take place:

> 'We have a game management plan. Wallabies are a big issue. They're picked off as they come forward (out of the bush). We shoot approximately 2000 a year. Fences don't work, maintenance is a problem and we haven't used 1080 for the past 30 years.'

Because some elements of the hunting fraternity can be a little lawless at times, engaging in trespass and damaging gates and fences, as well as illegally shooting wild animals and stock, many run owners cultivate their own group of recreational hunters, who are often given rough accommodation:

> 'We have regular shooters, that shoot wallabies for dog tucker. They spotlight and go on day shoots. That keeps the numbers under control.'

> 'We have shooters to shoot the possums. The farm supplies the bullets. They shoot hundreds of wallabies every year. One year, a group on one bush run shot 500–600 in an area that shares a Crown Land boundary.'

The shooters not only reduce the populations of problem animals, but also deter the activities of the lawless element, and, even, in one case, were said to pull out weeds. Even with the costs of maintaining rough accommodation, having recreational hunters control wild herbivore numbers may be a cheaper option than employing shooters or undertaking the shooting yourself. Of course, the recreational shooters do have a vested interest in the survival of the game species on the property, but most graziers appear happy to accept the presence of wild native herbivores in low numbers. Nevertheless, some graziers did not see recreational shooters as the whole answer to their game management problems:

> 'The wallabies have increased. Shooters don't come often enough, … they are only interested in stags.'

> 'Shooting wallabies is no incentive for shooters, other than dog meat.'

Because of its non-specific nature and perceived animal welfare problems, poisoning has become a controversial mammal control mechanism. This is especially the case with 1080, the elimination of the use of which is sufficiently popular among city folk to have become one of the targets in the outcome of the Tasmania Together community consultation process. Only five graziers said that they used 1080. They were concentrated in the 'high and moist country' enterprise group and tended to also state that they had gorse problems and used a combination of set stocking and rotational grazing. One grazier felt forced into it:

> 'Our biggest problem in managing stocking rates in the run country is game. It's bare as anything up there at the moment and it's not stocked. We have shooters but they're not making the impact they need to. It's not easy to shoot in rugged country. I hate to say it but it needs to be 1080ed. It's frustrating because we can't put sheep up there because it's been eaten out and that's no good for us and it's no good for the land either. If you have a bit of rain, you'll get a small amount of growth and they'll take it all. The roos etc. breed up due to the improved pasture and access to water. They come and graze the improved and go back to the hills.'

One grazier looked forward to its use in an adjacent property:

> 'Next door is planting timber which will be good because he will use 1080, which will account for a fair mob of wallabies.'

Two graziers explicitly stated that they did not use 1080. No other means of poisoning animals was mentioned by any of the 48 graziers.

A few graziers perceived that it was desirable to develop markets for wild animal products:

> 'Possums are a big problem. Last year and this year, we had a quota to kill for the American and Taiwan (meat) trade. When SARS hit, I think we shot 1200 possums, but the trade has stopped now. It had an impact for a short while, but not anymore. It's like the feral cat situation; they've all got babies.'

One suggested that there was a need to take advantage of a market for possum pelts:

> 'There is a market for possum and lamb skins. You can get $2–$5 for possum and $1 for a lambskin from NZ to make gloves. There's a possum fur trade in NZ.'

The sale of wallaby meat is another possibility, already realised on many properties, particularly at the pet food end of the market.

The carnivorous mammal that was most mentioned as a problem was the cat. The five graziers who mentioned this problem were all members of the 'improved country rotator' enterprise group, perhaps reflecting a concentration of mice in cropping country. Four of these graziers had observed an increase in cat numbers. The major problem that cats present to sheep is as a vector of toxoplasmosis. Feral cats have been ubiquitous in Tasmania since the nineteenth century, unlike the fox, which is currently struggling to establish, and which has the potential to be a major management problem in both the run and front country, unlike the cat.

Many graziers had observed a recent decline in Tasmanian devil numbers, undoubtedly due to the facial tumour disease that has been shown to reduce dense populations of devils to one-fifth of their previous numbers:

> 'We had devils, but have not seen any for a while. I found a dead one with a tumour on it. Were almost in plague proportions.'

Only one grazier said that he regarded devils as an issue, although, when their numbers were higher, poisoning of devils is known to have occurred on several properties. Devils can take lambs, but can also make a valuable contribution to the hygiene of properties by almost completely ingesting animals that have died from other causes:

> 'Devils were thick. They'd give lambs and young sheep a hard time.'

> 'I see a fair few devils. I still see plenty, they are about. They're a problem at lambing but I'd sooner have them than not. They will take the "poor twin" but that's nature as far as I'm concerned.'
>
> 'Dead lambs and sheep don't disappear like they used to.'
>
> '... sheep carcases – if devils are around, you're just left with a pile of wool.'

One grazier said that domestic dogs were a problem on his property, which may have been close to a town or city. Poorly controlled domestic dogs can form multi-breed packs that can kill or maim large numbers of sheep in a night. Working dogs on wool-growing properties are tightly controlled, and the pet dogs, if any, well contained.

White cockatoos, wood ducks and corellas were mentioned as problems by one to two graziers each:

> 'Cockatoos are a pest – there are 100s of them. You can get a permit to shoot four or five but all you're doing is to scare them on to the next farm. They're doubling their numbers all the time.'

It is unlikely that their status as a problem pertained to the economics of the run country. However, other graziers said that they valued the services, such as grub removal, that were provided by the avifauna:

> 'I used to poison rabbits, kangaroos, devils and cats, then we got swarms of plovers and native hens in the paddocks but we didn't have grass grubs.'

There was little mention of invertebrates as production problems in the runs. Grass grubs were noted by one grazier to be a problem, and, by two graziers, to be encouraged by fertiliser. One grazier noted a grasshopper problem during drought years. The corbie grubs and cockchafers that can cause chaos in improved pastures are native species that also occur in the runs, where they may benefit productivity by killing large rank tussocks, which are replaced by herbs and tenderer grasses. Plague locusts are not a major problem in Tasmania.

Weed management

A weed is a plant that is not wanted. In the context of wool-growing enterprises a weed in a run can be either a native, like prickly box, or an exotic, like gorse. Both are usually not wanted as it is perceived that, in abundance, they reduce wool production or the quality of the product:

> 'Gorse gets into the wool. You need to eradicate it, it contaminates wool.'

Some weeds, such as barley grass, can severely damage stock:

> 'Oh, it gets in their eyes and in their wool, in their skin. Very nasty.'

Table 2.7. Percentage frequency of comments on weeds

Gorse problem	58
Thistles an issue	27
Willow issue	23
Problem of weeds coming in with bought feed mentioned	19
Horehound an issue	17
Super advantages introduced species/weeds	10
Paterson's curse an issue	10
Ragwort problem	8
Hawthorn issue	8
Barley grass not good	8
Boxthorn problem	6
Vulpia issue	4
Heath developing in bush	4
Hakea a problem	4
Gorse has benefits	4
Gorse control successful	4
Fencing riparian causes bigger weed problem	4
Cumbungi issue (riparian/dams)	4
Capeweed issue	4
Broom an issue	4
Broadleaf weeds an issue	4
Blackberry on creeks	4
Removing saggs	2
Willow present but not considered problem	2
Weeds take off after drought/bare ground	2
Weed control requires additional/unavailable labour	2
Water lilies (riparian) issue	2
Wandering Jew present	2
Twitch an issue	2
Sycamores present	2
Slender thistle an issue	2
Silver grass issue	2
Serrated tussock a problem	2
Removing bursaria	2
Prickly box issue	2
No invasive weeds in bush	2
Invasives in bush	2
Hawthorn not an issue	2
Foxglove present	2
Fire necessary to control heathy plants in grassland	2
Elderberry present	2
Dog's tail	2
Cotoneaster present	2
Bracken a problem	2

Gorse has well and truly escaped from the hedges described by Louisa Anne Meredith in the 1840s. It was by far the most prevalent weed problem among the 48 graziers (Table 2.7):

> 'Gorse is the most obvious weed, it's uncontrollable.'

Problems with gorse were mentioned by graziers in all enterprise groups, with the 'improved country set stockers' and the 'high and moist country graziers' having the most concern (refer to Table 2.1). Gorse can form impenetrable thickets that suppress grass growth and shelter wild animals (see Figure 2.12). Mature bushes will be grazed by sheep, but only in desperation, producing topiarised bushes to their head height (as shown in Figure 2.13). Its provision of shelter, and role as source of fodder, are the reasons why some graziers said that gorse had some benefits:

> 'The run to the south is a gorse bank up to the plateau. We put lambing ewes in there for shelter.'

> 'Gorse is left on the stony ridges. It's good sheep feed if you keep it burnt.'

> 'If you don't have gorse, you'd need a tree break.'

However, most graziers put resources into the control of the species:

> 'We spent $700 to spray the gorse in a small area, plus labour costs.'

Figure 2.12. An impenetrable mass of gorse (*Ulex europaeus*), recently sprayed (Jamie Kirkpatrick).

Figure 2.13. Gorse topiarisation (Jamie Kirkpatrick).

'I allow $10 000 a year for gorse control not including labour.'

'Gorse is the main problem on the runs, that and cotton thistles. We sprayed the gorse with a helicopter and that did well. Now we do follow-up sprays, spot spraying to keep it under control. We only have patches of gorse here, not huge areas but the regrowth is expected for many years.'

Two graziers said that they had succeeded in getting rid of gorse from their runs:

'Gorse was a huge mess 50 years ago. Gradually there's not much left.'

This is a major achievement that requires persistence in locating and eliminating spot infestations caused by bird or water transport of seed, or by germination from soil-stored seed. The herbicides that are most effective in killing gorse are expensive, well beyond the monetary returns from increased fodder production in runs. Burning will set gorse back to ground level (see Figure 2.14), but it sprouts from base and roots after fire, and the germination of its hard-coated seeds is stimulated. Nevertheless, burning does allow grasses to be grown for a few years where once there was only gorse, and makes gorse more palatable. One grazier, who had inherited dense gorse as a result of years of neglect by the previous owner, utilised a steamroller to crush gorse before burning and spraying. Another grazier had a novel solution to the gorse problem:

'We should call gorse cultural heritage and it would be gone tomorrow.'

Figure 2.14. Burning gorse (Kerry Bridle).

Biological control has had some effect on gorse:

> 'The gorse mite was effective about two years ago. It stopped flowering.'

However:

> 'We've got about 200 ha of gorse, the Jericho rose we call it. We had the gorse mite which made it easier to manage and then the apple mite was introduced and it ate the gorse mite.'

Graziers who said that they had a gorse problem also tended to have said that the bush was a source of feral animals and that they had problems with tree dieback, willows and Paterson's curse (*Echium plantagineum*). They also tended to say that they used 1080, perhaps a reflection of the difficulties of wild animal control by shooting in gorse-infested landscapes.

Apart from gorse, hawthorn (*Crataegus monogyna*) and South African boxthorn (*Lycium ferocissimum*) are the only two shrubs, both introduced for hedging in the first decades of sheep farming in Tasmania (see Chapter 1, Figure 1.6), that were mentioned by more than one or two graziers as a problem (see Table 2.7). Both these species are prickly, have fleshy fruit, and seed spread by birds. However, the hawthorn hedges of the country settled before 1850 are too culturally important to allow their elimination, although some desire it:

> 'These hawthorn hedges were put in as windbreaks in years gone by, now we've lost the lovely English touch of cutting them every year. We've allowed them to become Australian.'

Others see no problem:

> 'Hawthorn's not a problem. Trim hedges and it would make a difference. You need to let the stock in to keep it under control.'

The rotational enterprise group most mentioned boxthorn as a problem. No-one seems to have a cultural attachment to this species, which is well adapted to dry ground:

> 'Started in hedgerows and has spread over the hills now – five years ago not the numbers there are now – very difficult to get rid of.'

Broom (*Cytisus scoparius*) was the only other exotic shrub mentioned as a problem by graziers, both in the high and moist country enterprise group, as were the two graziers that mentioned the native hakea (*Hakea microcarpa*) as a problem. Two in the 'improved country rotator' enterprise group mentioned the native prickly box as a problem. The exotic woody scrambler, blackberry (*Rubus fruticosus*) and several exotic shrubs that can be weeds in the bush were mentioned, but not as economic issues.

Willows were mentioned as an issue by almost one-quarter of the 48 graziers. Willows have been regarded to have a valuable role in stabilising river banks:

> 'I like willows. There'd be an erosion problem on the river without them. I don't call willow a weed.'

At the same time willows can render some of the best potential pasture useless, and consume large amounts of scarce water, which could otherwise be used for irrigating crops. It is therefore not surprising that those who said that willows were an issue also tended to mention that they cropped. Cumbungi (*Typha* spp.) was mentioned as an aquatic weed by some graziers. It can choke up dams and slow waterways.

Thistles and horehound (*Marrubium vulgare*) were the herbaceous weed species most mentioned as a management issue (see Table 2.7). They both tend to be most successful in their establishment on bare ground and were conjointly mentioned by graziers significantly more than could be expected by chance. Horehound was most mentioned by the graziers in the 'improved country set stocker' enterprise group, while thistles were commonly mentioned by both graziers in this group and the 'improved country rotators' group (see Table 2.1). The properties of the graziers who mentioned a horehound issue had significantly lower mean annual rainfall (512 mm) than those graziers who did not (615 mm). Two other herbaceous weeds that colonise bare ground,

Paterson's curse and ragwort (*Senecio jacobaea*), were preferentially mentioned by graziers in the 'high and moist country' enterprise group, reflecting their need for high rainfall. Capeweed (*Arctotheca calendula*) and other broad-leaved weeds were mentioned as problems by some graziers, largely in the 'improved country rotator' group. The direct role of bare ground, and the influence of drought, in causing infestations of all the above herbaceous weeds were noted by some graziers:

> 'You get a lot of horehound if you overgraze in a drought.'

Bare ground also favours the establishment of annual grasses such as *Vulpia* and silver grass (*Aira* spp.) that are regarded as undesirable components of native pasture.

Control of weed species can be partially achieved by management of stock (Mokany *et al.*, 2006). Although each type of stock management does tend to favour different weeds, bare ground can be minimised under any grazing system, and this action alone can achieve prophylactic outcomes. Strategic grazing and burning can also be effective in controlling some weeds. Herbicide application and mechanical removal are expensive options, usually adopted in extreme situations, such as those presented by years of neglect:

> 'From the Depression until post-war there was little property management such as weeds, fences. Farms went downhill and gorse got away.'

Many graziers saw a relationship between fertilisation and weed problems in native pastures:

> 'You need to look at the costs. If you put super on the end runs, it encourages the weeds. You can increase your stocking rate and grubs and you get more problems.'

> 'There's a cost associated with fertilising the semi-improved country. It was originally sown but the species have gone and we've got a lot of broadleaf weeds, and twitch. Low quality feed and not enough of it to warrant fertilising it. It needs to be aerially sown again.'

The problem of introduction of weeds in stock feed was well recognised:

> 'We're conscious of weeds in feed. We have a week of hand weeding after feeding.'

> 'Weeds turn up after a drought. You buy grain and we got Paterson's curse everywhere. They say they quarantine the grain now but … . We haven't seen it again. The climate doesn't suit it and the sheep eat it.'

> 'We supplement feed local grain to ewes. The mainland feed is expensive and brings in weeds.'

Conclusions

The natural environment of the individual property strongly influences the nature of the wool-growing enterprise. This is indicated clearly in the environmental distinctiveness of most of the groups in the enterprise classification, a classification that did not use environmental variables as inputs. This environmental distinctiveness not only pertains to rainfall and altitude, but also to the proportions of improved and bush country (refer to Table 2.2), which are strongly likely to reflect land capability. Different environments present different management problems, particularly with diseases, pests and weeds, and different production opportunities, such as those related to kangaroo grass, tussock grass and sagg. To a large degree, graziers seem to have adjusted their activities to the environments of the properties on which they work their sheep.

Despite the importance of environmental influences, there is no doubt that different graziers have developed very different systems to produce wool from runs on properties with highly similar natural environments. The 'improved country set stocker' and 'improved country rotator' enterprise groups do not occur in distinct natural environments (see Table 2.2), but rather are a product of human choice. There may be poor or good economic choices in the management of runs for profit, but there is certainly no one right answer. The choices involved in deciding such things as whether or not to fertilise, or whether to set stock, rotationally stock or cell graze, can be influenced by economic circumstances, variation in the preparedness, or need, to take economic risks, and the weight that is placed on the non-economic aspects of farm life. Wool growing in Tasmania is still largely the province of families, many of whom have lived on the same land for many generations. People become attached to places, environments and ways of doing things, and weigh these attachments against whatever appears at the time to be economic rationality. The runs are usually not just money machines, but are also valued for themselves. The next chapter explores attachments to the nature of the run country and how they express themselves in the conservation management of wool-growing properties.

Notes

1 Comments made by the graziers were converted into qualitative variables. For example, if a grazier commented upon the management problem presented by gorse (*Ulex europaeus*) they scored 1 on the variable 'gorse problem'. If they did not comment on a gorse problem they scored 0. This process enabled an exploration of the relationships between pairs of comments using a statistical procedure called Chi-squared. This procedure allows the determination of the probability that two independent comments were made by the same graziers more or less than expected. Because a formal questionnaire was not used, the negative (less than expected) results are not as reliable as the positive (more than expected) results. If we write that those who said that they practice rotational grazing tended to also say that they had recent management changes, this means that the same people made these comments more than would be expected by chance. It does not mean that all people who said that they rotationally grazed said that they had recent changes in management or vice-versa, only that more than could be expected by chance did so. Chi-squared cannot be

used if expected values are low, so this form of analysis could only be used for the more common comments. Less common comments are included in the tables that accompany each topic along with the more common comments. These tables show the percentage of graziers who made a particular comment. It needs to be emphasised that the presence, or absence, of a comment does not necessarily translate into the presence or absence of an action or opinion, partly because the discussions were informal, with no structured set of particular questions, and partly because of the tendency of human beings to moderate their conversations to be considerate of the opinions and feelings that they feel may be held by those with whom they are conversing. The qualitative data relevant to the enterprise were used to produce a classification of the observations, experiences, attitudes, management perspectives and management procedures of the graziers who managed the 48 properties. This was done by first using a program that does one to four dimensional ordering of individuals (the properties), such that properties that have similar sets of comments have similar scores. The procedure is called multidimensional scaling and was undertaken in a program called DECODA using the default options (Minchin 2001). The scores for each of the four dimensions were used as the input to an agglomerative (grouping from below) classification using Ward's method of averaging distances with each fusion and Euclidean distance. This was done in the statistical program MINITAB (Minitab Inc. 1998). The quantitative data for rainfall, altitude, property size, the proportions of the property consisting of cropland, improved pasture, native pasture and native bush, and the stocking rates on different types of pasture were related to the qualitative variables using analysis of variance (ANOVA), a technique that allows the calculation of the probability that a quantitative variable varies significantly between classes. For example, it was possible to test whether graziers who commented that 'dieback is an issue' had properties in areas with higher or lower rainfall than those who did not make this comment.

2 The extent to which the 48 properties typify wool-growing enterprises in Tasmania can be determined to some degree by comparing attributes of property size with 2004 data from ABARE. The mean area operated by a sample of mixed enterprise sheep farms was 1002 ha, while for specialist sheep farms it was 1665 ha. Data on property size were also obtained by Bob Reid and Associates (2003) for the 8 x 5 program, which focuses on increasing profits from wool production. These data were obtained for 23 properties in 2000/2001 and 24 properties in 2001/2002. The participants in 8 x 5 were volunteers, many re-establishing flock age structures after some bad drought years. The 8 x 5 mean property size for 2000/2001 was 2246 ha. In 2001/2002 it was 1895 ha. The range in property sizes was from 314 to 6657 ha. The 48 properties had a mean area of 4032 ha, a median area of 2175 ha and a range from 340 to 18 242 ha. The mean percentage of 8 x 5 properties under crops and improved pasture was 47.4 for 2000/2001 and 51.2 for 2001/2002. The 48 properties had a lower mean, 34.37 per cent, with a range from 0 to 92 per cent, reflecting the focus of the discussions on graziers who had native vegetation.

3 $P < 0.05$ using ANOVA or Kruskall-Wallis tests. Pearson's correlation coefficients between return on investment in wool production and the percentage of the property consisting of native pasture were also not significant. Attempts to fit quadratic and polynomial curves to the relationship between the absolute values for these two variables also came up with levels of explanation close to zero.

4 $P < 0.05$ using Pearson's correlation coefficient.

Chapter 3

Conserving on the run country

J.B. Kirkpatrick, K.L. Bridle, J. Edwards and J. Vercoe

'There's a fine line between being driven for profit and looking after your country.'

Summary

Graziers are aware of the importance of maintaining landscape productivity, mainly through preventing both accelerated soil erosion and salinisation. Many look after their streams, usually in cooperative endeavours. Most graziers value nature on their runs: the sense of belonging; the beauty of trees; the birds; the non-herbivorous mammals. They are distressed by widespread tree dieback and commonly fence off bush or streams and covenant parts of their runs.

Introduction

Conservation is defined as the wise use of resources, which begs the question of the nature of wisdom. This chapter uses a definition of conservation based on the general social consensus – that it is unwise to permanently reduce the productive capability of land and water, or to render any native organisms, or the communities they form, extinct.

The chapter is largely based on reporting and analysis of the discussions with 48 managers of wool-growing properties (see Table 3.1). Methods are described in detail in Chapter 2. It also utilises data from more formal questionnaire surveys of 400 randomly selected landowners, conducted by the Australian Bureau of Statistics (ABS) in 1992, for Greening Australia, and, in 1998, for Edwards and Kirkpatrick as part of a project funded by the precursor of Land and Water Australia (Edwards 1998). These surveys covered farmers in the northern and southern Midlands and the Central

Table 3.1. Percentage frequency of responses related to nature in the runs by enterprise groups
Responses that occur in at least 20 per cent of the enterprises in at least one group are shown. The highest percentage frequency for each comment, and values of 50 and above, are shown in bold.

	1	**2**	**3**	**4**
Like trees/bush – aesthetics	**67**	37	33	36
Regenerating parts of bush	**50**	37	33	18
NFS scheme	**42**	12	0	0
Fenced off some bush/remnants	**42**	37	22	9
River not fenced	**33**	6	0	18
Stock access to riparian area not a problem	**25**	6	0	0
Burn saggs	**25**	0	11	0
Wombats present	**25**	0	11	9
Environmental accreditation valuable	**25**	12	22	0
Dieback an issue	**50**	**50**	11	18
Values associated birdlife	8	**25**	22	18
River/creek frontage present	25	12	**44**	9
Decrease in devil population	25	19	**33**	27
Thickening of woody vegetation a problem	17	0	**33**	18
Increase in possums	8	6	**33**	9
Increase in cats	0	6	**33**	0
Have whole farm plan	0	12	**33**	18
Manage parts of bush for conservation	8	6	**22**	0
Accreditation criteria difficult to develop	8	0	**22**	0
Poa present	17	0	**22**	9
Dieback cause by combination of factors	8	12	**22**	9
Tussock present	8	6	**22**	9
Bandicoots present	0	6	**22**	0
Brown tussock moth present	0	6	**22**	0
Kangaroo grass present	17	19	0	**64**
Wallaby grass present	25	19	22	**36**
Covenant on area/some of area	8	19	11	**27**
Silver tussock present	8	12	22	**27**
Part of river/waterway fenced for natural regeneration	8	12	22	**27**

1 = 'improved country set stockers', 2 = 'improved country rotators', 3 = 'high and moist country graziers', 4 = 'warm country graziers'.

Highlands. Another source of information is the work of Vercoe (2003) who used formal questionnaires, qualitative interviews and focus groups to assess the problems and achievements of the Private Forest Reserve Program (PFRP), a program that has involved a large number of wool growers.

The major aim of this chapter is to document the ways that graziers think about and conserve nature in their run country. Such conservation includes protection of soils and water, as well as the maintenance of natural landscapes, native species and native communities.

Maintaining soils

> 'To grow the premium wool we need to keep the country at a sustainable level. We realise that we can't push it and we're not trying to. We will increase our return by improving the quality of our wool. The only way to make money is to grow the best, to make the best.'

There are two major threats to the long-term productivity of the soils of the run country: accelerated soil erosion and dryland salinity. Naturally saline soils occur in the run country (see, for example, Figure 1.23), but are unlikely to spread unless the hydrology of their catchments is altered in a manner that increases ground water flows from the sandstone that is the major source of salt. This could result from massive tree clearance on sandstone on properties upstream from the saline soils on the runs. The importance of trees in preventing salinity was noted by three of the 48 graziers:

> 'I'm sure it's useful to have trees on slopes as a salting deterrent. Trees are valuable, if you don't have them might be problems down below.'

One grazier stated that they kept the *Poa* tussocks on naturally salty places (see Table 3.2). Another said: 'We cannot discount the importance of native perennial grasses and shrubs' ability to hold and use the water in the topsoils of the upper catchments if they are managed well', reflecting an understanding that perennial vegetation cover as a whole is important in preventing problems with salt.

Potential problems with salt are not as widespread in the Tasmanian runs as potential problems with soil erosion. Accelerated soil erosion is reversible, but only over timescales of millennia, because of the slow rate of weathering of bedrock in temperate climatic conditions. The most fertile upper part of the profile tends to be the first eroded. Accelerated erosion will not occur if the natural degree of protection given to the soil by vegetation and litter is maintained. Unnaturally bare ground is a recipe for accelerated erosion on all but the gentlest of slopes, which are the places where topsoil eroded from the upper slopes often ceases its journey: 'For soil conservation you need the most productive grasses – and don't allow bare spots and erosion.' The maintenance of ground cover is not usually difficult in the runs, making wool growing in these areas potentially one of the few physically sustainable agricultural activities on sloping ground. Several graziers commented on the importance of maintaining ground cover, with one stating that their first priority was looking after the soil (see Table 3.2).

Some graziers mentioned particular erosion problems. One grazier from a high altitude property mentioned freeze–thaw erosion. Another avoided using cattle on

Table 3.2. Percentage frequency of comments related to soil conservation

NFS scheme	15
Have whole farm plan	15
Don't have whole farm plan	8
Avoid dusty conditions/maintain ground cover	8
Trees reduce salinity	6
Fenced off some hilltop	6
NFS scheme well targeted	4
Fenced hill to reduce erosion	4
Bush reduces erosion	4
Wombats potential erosion problem	2
Want NRM plan	2
Salt areas maintained as *Poa* tussock	2
Rotating maintains biomass/reduces erosion	2
Priority is looking after soil	2
Potential erosion problems on sweeter country	2
Freeze–thaw erosion an issue	2
Fire breaks cause erosion/stopped making them	2
Erosion problems with cattle in bush	2
Destock in dry years	2
Better ground cover makes rainfall more effective	2

their runs because they thought that they caused erosion. A third stopped making firebreaks because of erosion problems. A fourth was aware of the potential for tunnel erosion on particular soils: '(Our) native grasslands are on mudstone/sandstone, light sandy or silty soils, generally on tunnel erosion areas.' A fifth did not cultivate their silver tussock flats because of potential flood erosion:

> 'Silver tussock paddocks, when underwater, are too risky. I'm not prepared to take risks, the land is too low, we might lose it. I watch the neighbour's soil drift down, he has big soil erosion issues.'

The major cause of bare ground in the run country has been the strong tendency of sheep to concentrate their grazing on the warmest slopes, those facing west, northwest and north. Leonard and Kirkpatrick (2004) found that sheep droppings were concentrated on such slopes on dolerite soils, and that the surface soils on the north-facing slopes were red, compared to the dark surface soils elsewhere, a possible indication of a loss of the organic-rich surface horizon typical of soils under native grasslands on this rock type. Steps, apparently sheep paths, but actually mass movement features called terracettes that are induced by overgrazing, are common features of the north-facing slopes (see Figure 3.1). These north-facing slopes tend to erode least where paddocks do not include other types of country. Most of the damage occurs in

Figure 3.1. Terracettes and slumps on a north-facing slope (Jamie Kirkpatrick).

topographically heterogeneous paddocks, with bare ground being created on north-facing slopes, while forage remains uneaten elsewhere. The increased set stocking rates that result from fertilisation of the runs appear to exacerbate the problem, as is indicated by the strong tendencies of those graziers who said that they participated in the North-Facing Slopes Program (NFS) to belong to the 'improved country set stocker' enterprise group (refer to Table 3.1), and to say that they fertilised their runs. They also tended to say that they were regenerating part of their bush; had tree dieback problems; valued saggs and tussocks for shelter for both off-shears and lambs; and, surprisingly, did not fence their waterways. None of the graziers from the 'high and moist country' and the 'warm country' enterprise groups said that they participated in NFS. This is more likely to be related to the high rainfall on their properties than to stocking regimes, which are highly variable within both of these groups. Few in the rotational grazer enterprise group participated in the NFS, probably because rotational grazing, in itself, is a solution to the problem:

> 'The she-oak hill was red, grazed to bare earth. With a rotational grazing system it now has 100 per cent cover.'

The main mechanism used in the NFS was to fence paddocks according to the carrying capacity of the land, based on soils, geology and erosion potential, forcing the sheep to eat forage they could previously ignore, and allowing the adjustment of stocking rates to environmental conditions. Some of the worst-eroded slopes have been fenced and have had trees planted on them. The program was highly successful, and was perceived as well targeted by at least two graziers (see Table 3.2).

Whole farm planning, now renamed 'property planning', involves designing land use to suit productive potential, among other aims. Most of the graziers who said that they had such plans were in the 'high and moist country' enterprise group (see Table 3.1), and they strongly tended to also say that they rested runs in spring. Whole farm planning was very popular in the early 1990s, when 45.6 per cent of the respondents in the ABS survey had such a plan. In the repeat survey in 1998 only 36.9 per cent of respondents had one. However, property plans are now becoming more popular:

> 'We are actively getting a property plan that includes production and the environment. We didn't want just a biodiversity plan. The environmental part of the production stuff needs to be in place. It's the plan forcing us to focus – to get some time lines together – how we can fit grazing management into native pastures, remnants, rare and threatened species. We have focus in our heads and I'm sure we're getting 80–90 per cent of it right – the planning process has forced us to get a bit focused. We've developed up an A3 spreadsheet for 52 weeks of the year. We blank out when we shouldn't be grazing (a particular paddock) for the production areas as well as for the rare species. We will work the grazing management around it.'

Planning for all aspects of sustainability does not have to take place within this formal framework:

> 'We have no farm plan at the moment. I work on an old method. My gut feeling. If it aches I know I've got to do something, if it doesn't I don't. Thinking through your gut is better than thinking on paper.'

> 'I wouldn't be interested in the farm that consisted of green paddocks with four fences around them. It's of no interest to me at all. I like living in or near the bush, that's basically what is important to me. And because I miss this country, and am responsible for its management, I take it seriously. Grazing is only part of the management, so, fire, and weeds, and erosion, and everything and anything else that happens in the bush, whether it is firewood collection, or mineral extraction, or logging, or just botanical species composition. All these things matter to me. So it is all paramount in my consideration when I'm managing. I will just address this year, some of this country that was ringbarked many decades ago and is still open, has got no tree cover at all, and it had been, well I think the people before stocked it really heavily, and I still stock it moderately. I had a couple of botanists from the Non-forest Vegetation Program, they did do a botanical survey, but it was springtime, and the sheep were off so it was a good time to see what was there. And after they had looked around them, they had a page of species. And I said, if this was still forest, and it had its original eucalypts – *amygdalina* and *globulus* overstorey – is there anything there

that is missing. And they said, no, the full range of species that we would expect to collect from a site such as this are present. And I think they only collected four weeds. So that was interesting, and it was pleasing to me. So as that country grows back into bush I think it will recover most of its natural values as well ... the botanists who did the assessment for the program described my run country as above average. I can't remember the exact words, but something like the diversity or the health of the bush, because it is well managed. So that sort of qualifies my own value judgements on the country. I just feel that country will stand sheep grazing, so I have done that sensibly and lightly.'

Maintaining water quality and quantity

A large proportion of the properties owned by the 48 graziers had river or creek frontage, although not all said that they did (see Table 3.3), the comment being most frequent for graziers in the 'high and moist country' and the 'improved country set stocker' enterprise groups (refer to Table 3.1). Most land titles in Tasmanian grazing country incorporate streams and other water bodies, so there is no obligation to fence so that streams and wetlands are separated from the rest of the property.

The benefits of fencing streams, and removing exotic plant species, are as much social and environmental as individual:

> 'We removed the willows because we want to get the native vegetation back along the river, for river health and shelter for fish.'

Stock can increase turbidity by causing accelerated erosion of banks and can pollute the stream with their faeces and, occasionally, their corpses. These, of course, are largely problems for the landowners downstream, as are the many weed species swept downstream on the currents. However, fencing does have economic benefits:

> 'Fencing removes time looking for stock in the river; it used to take two hours a day in summer.'

A lack of fencing and weed control along streams has led to an almost total destruction of the native riparian vegetation in the longest settled areas (Daley and Kirkpatrick 2005). The contrast between the original and current riparian vegetation is extreme as shown in Figures 3.2 and 3.3.

The restoration of river health requires cooperation between landowners in catchments:

> 'We can't solve the willow problem on our own.'

There is no point in fencing your side of a stream if stock can wade across, or downstream, from a neighbouring property:

Table 3.3. Percentage frequency of comments related to water bodies and streams

Willow issue (riparian)	23
Irrigating	23
River/creek frontage present	21
Part of river/waterway fenced for natural regeneration	17
River not fenced	15
Stock access to riparian area not a problem	8
Decrease in waterbird population	6
Fencing riparian causes bigger weed problem	4
Cumbungi issue (riparian/dams)	4
Blackberry on creeks	4
Stock shouldn't always have access to river	4
Stock access to river limited by scrub	4
River fenced	4
Willow present but not considered problem	2
Water lilies (riparian) issue	2
Wetlands not managed differently to other bush	2
Water an issue in runs	2
Want to put in troughs to prevent riparian degradation	2
Want to prevent stock accessing riparian area	2
Plantations in upper catchment reduce water flow/availability	2
Periodic problems with salinity of irrigation water	2
Periodic problems with salinity of stock water	2
Fencing improves water quality	2
Flooding an issue	2
Catchments drying due to increased perennial grass cover	2

> 'Fencing along the river is new. Stock are only in there a couple of days a year. Our neighbours are "less religious" about maintenance.'

A certain degree of altruism is also required, given the expenses of fencing and weed control, not all of which can be recouped from the Natural Heritage Trust. Such cooperative altruism has been much in evidence among a subset of graziers, such as many of those along the upper Macquarie River:

> 'On our river frontage, we have a mile on both sides, there's a nice little hakea. There is some native vegetation but unfortunately there's more willow and gorse than anything else, and cumbungi. This was a major worry some time ago. Our Landcare group got some grant money to clear out the river. There are still willows but it's not clogged. The advocates of fencing don't know that it causes a bigger weed problem, even hawthorn on the flats. Grazing is the only weed control used but it's not necessarily effective. I haven't done any spraying.'

> 'I encourage birds. They give me a lot of pleasure. I'd hate to see the place denuded of trees.'

> 'Biodiversity is a garden. You can see and hear different birds. In the middle of Victoria all you hear are starlings, sparrows and swallows.'

> 'In another life, I'd be a bird watcher. I really like native birds, for pasture pests you need lots of birds around. The grubs here only affect small areas. It's important to keep lots of birds around, they eat a lot of grubs and I can't see that they do much harm. To have birds you need trees and understorey.'

Graziers who said that they irrigated also tended to say that they valued birdlife, perhaps because of the opportunities they had to observe waterbirds, or perhaps because they had learned about integrated pest management. Many of the graziers valued the bush for its ecological processes, biodiversity in general, or as habitat for animals (see Appendix 2).

Recreational values of the runs were mentioned by several graziers:

> 'Our international visitors love it. People from Germany go bananas and the interstaters like it.'

> 'I'm putting a bird sanctuary on the dam area. I saw 1500 ducks down there. It's the only place in Tassie where you can pull up and watch birds on a bitumen road. People stop there every day. We had some visitors from Canada (in the tourist cottage) and they booked in for a night and stayed for five weeks. She identified 60 different birds on the place and she never got a native hen or a wood pigeon.'

One grazier mentioned the importance of having bush for their children, while another valued their bush for its wilderness qualities. Some were fond of the bush because it reminded them of their younger years:

> 'I don't mind the bush. It's nice to have a bit of bush. As kids we grew up in the bush, hunting and shooting.'

In an earlier study of the attitudes of Tasmanian landowners and managers to their native grasslands (Gilfedder and Kirkpatrick 1995a, 1997), about half of the graziers in the sample valued lifestyle benefits and aesthetic qualities, with about a quarter valuing their native grassland as an example of the original vegetation type, for its biodiversity and for its potential for future genetic material. In the 1998 ABS survey, respondents indicated that they believed the main benefit of native vegetation was the provision of shade and shelter for stock, with the next greatest benefit of native vegetation being the protection of biodiversity and conservation values.

Observations on trees

Many graziers saw the dieback of trees in their runs (see Figures 3.5 and 3.6) as an issue (refer to Appendix 2). This tree loss has been an ongoing concern. The ABS surveys found that almost 60 per cent of respondents in both years had observed tree decline on their properties. Between 1992 and 1998 the percentage of respondents noticing over 500 hectares of their property affected by tree decline increased from 6.7 to 20.1, indicating a marked increase in its severity where it occurred (see Figure 3.7). Three graziers noted that white gums were particularly susceptible to dieback (Appendix 2).

The graziers who saw tree dieback as an issue had properties with significantly lower mean proportions of improved pasture than those who did not (19.12 per cent compared to 44.62 per cent). They also had properties with significantly lower mean annual rainfall than those who did not (541 mm compared to 630 mm). This latter relationship suggests that drought is the major reason for tree dieback, a point not missed by most of those graziers who commented on its causes:

> 'Dieback is caused by old age predominantly, followed by drought. Stock and possums also have an impact, in equal number. Possum populations have increased dramatically, thanks to the farmers who provide them with plenty of food and water, and enable them to breed more than once a year. … but you can see, every tree, live and dead, has at least one possum family living on it.'

> 'Dieback. Is it old age? Rainfall? Not fertiliser because they're dying in the run country too. We've had dry seasons for a long time, there's no water underneath. They're sick and the possums come.'

> 'Dieback is caused by drought, possums and insects.'

> 'I am concerned about tree decline. I think possums have a major impact.'

> 'The cause of dieback is old age, just a shell on the outside and inside are rotten possums. Super has a little bit to do with it and drought.'

> 'The big old gums are more susceptible where they're sparser. White gums are susceptible to cultivation. Black ones are deeper rooted.'

Almost half as many graziers found woody vegetation thickening a problem as found dieback to be an issue (Appendix 2). They were concentrated in the 'high and moist country' enterprise group (see Table 3.1). The properties managed by the graziers in this group may not have been subject to the same drought stresses as those in other groups because of their high rainfall (see Chapter 2, Table 2.2). A major worry was the thickening of wattles:

> 'Under set stocking, the young wattles and things get nipped off at the young stage. One bite and that's it. But with the cell grazing, the wattles can get up to quite a height. They have decent size stems, and it means

Figure 3.5. Dieback of trees at the margin of a forest remnant (Jamie Kirkpatrick).

Figure 3.6. Dead cider gum near Liaweenee (Jamie Kirkpatrick).

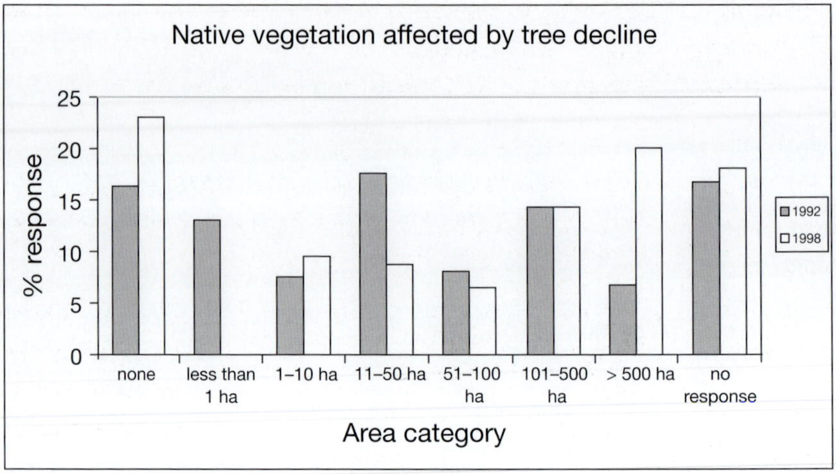

Figure 3.7. Percentage of respondents to ABS surveys who had different areas of tree dieback on their properties in 1992 (left bar) and 1998 (right bar).

> that they have got some reserves, so if they get grazed off, they keep coming back. So maybe we might have to go back to fire again, if that looks like getting out of control, we could use fire as a way of stopping that getting out of control. But it's all happened because all that country was rung out (ringbarked), at the end of last century (end of the 1800s) and after that there was enormous regrowth, and so the eucalypts were all overstocked – too dense. And so, basically, we've been there for 30 years, and the eucalypts haven't grown in 30 years. And now they're dying. The eucalypts just stalled, because they got too big, and now because of drought, there is enormous dieback. And now the cycle's starting again, with the wattles coming in. I think that it's actually starting now the cycle it always had. When they rung it out they broke the cycle, and so it's taken this time now for the wheel to start turning again.'

A few graziers observed that hot fire can kill or damage trees:

> 'About two-thirds of the property is dolerite, and the other third is partially alluvium and the rest is sand, and the sandstone country is much less fertile, and, if it is ploughed, sown and fertilised, it will grow grass, but the bush will only grow sheep if it is burnt regularly. And if it is unburned, a lot of it will grow stringybark, which doesn't like fire. I should qualify that by saying the timber is damaged more easily by fire because the fibrous bark burns, so it is better management to try and keep fire away from the stringybark forest if you're trying to manage it for timber.'

Another noted that cool fires did not promote eucalypt regeneration. A high country grazier had observed that the cider gum (*Eucalyptus gunnii*) established best among the protection of the prickly shrub, *Hakea microcarpa*.

Observations on other native plant species

Graziers had a high degree of awareness of the major groups of native grasses, particularly those in the 'warm country' enterprise group (see Table 3.1). The genus *Poa* was mentioned most, largely masquerading under a variety of common names: silver tussock, tussock and blue grass tussock (See Appendix 2). Kangaroo grass (*Themeda*) and wallaby grass (*Danthonia*, now *Austrodanthonia*) also had a high incidence of mentions (see Appendix 2).

Observations of species of little or no importance for forage were limited. One grazier referred to a threatened species, the Tunbridge buttercup (*Ranunculus prasinus*) on their property:

> 'There's only one patch of some of these things, and I quite like to see the trees and the orchids, and the Tunbridge buttercup, but I'm not particularly interested in that, give it or take it. It's not my type of flower.'

One grazier noted the importance of riparian vegetation for rare plant species (Appendix 2). Many graziers highly appreciated the wildflowers in their runs, without necessarily knowing their common or scientific names:

> 'Bush patch is locked up, 40 acres, for flowers, orchids and blue pin cushion.'

> 'We'd much rather go in the bush than paddocks, it's more interesting. There are different flowers, orchids. We can't clear it even if we wanted to.'

Vases containing such flowers or watercolours of them painted by family members were seen in several houses. Some graziers propagate and grow local native plants in part or all of their home gardens.

Observations on animals

Most graziers seemed to have some degree of empathy with their sheep, several interpreting their emotions and desires:

> 'It (rotational grazing) was confusing, for the sheep more than anything. If they would have been left alone set stocked doing their own thing they probably would have been fine, but because they were grouped up and being shifted every few weeks at a time when there wasn't much feed, it was a bit tough on them.'

> 'They get to know their own run, wethers, and when you take them away from it they quite often want to get back there too.'

> 'The older wethers like this country better than the younger wethers. They can fossick around here all day long. They'd rather be out here than in the grassy paddocks.'

Graziers also had attachments to native animals, occasionally even ones that interfered with the economics of their enterprises:

> 'We have wallabies, possums, wombats and incidental foresters. When we built here in 1978, as young fellows, we'd shoot 300 wallabies a night just on the flats when the property was all bush. There was more game then. Now numbers are building up. I see wallabies in the middle of the day in paddocks with no feed. I don't like shooting them, I really like them. I don't like possums; I shoot one a night if they come into the garden.'

> 'I have 1400 acres that's not good for grazing. It's OK for logging except for eagles' nests (two)… . I would prefer that the eagles' nests were on the neighbour's property! I like to see them, and I won't shoot them, but I am restricted by them being there.'

One property owner looked after orphaned wombats. However, no-one expressed an emotional attachment to possums, and many conservation-minded graziers disliked all pestiferous wild animals:

> 'In the improved runs, in a couple of places, there is dieback in the white gums. I have fenced off the gullies. They're damper and we get regeneration in there. I'd like to do a lot more, even only 5–10 acres. In future years we may be like the Midlands, bare hills. I'd love to be able to fence off the gullies more, fencing little creeks and streams. They're havens for pest animals such as wallabies and possums. I don't like them, that's the downside.'

A substantial number of graziers had seen changes in wild animal populations (see Appendix 2), including the rise and crash of the devil population discussed in the previous chapter. These observations were most common among graziers in the 'high and moist country' enterprise group (see Table 3.1). On some properties many changes in animal populations had been observed:

> 'The animals certainly rise and fall. The wallaby numbers are the obvious ones. The drier summers, and colder wetter winters, there is more pressure on them in the bush, and they come onto the open country more, we see them more. … I have seen more wombats coming out into the paddocks. They tend to be the older, mangier animals. Devils are the big question. Devils tend to come out after midnight. That is why we tend not to see them, we tend to be asleep. But the last two or three years, I

haven't seen the devil on the farm, and I'm very observant and I have been watching for them, but the last couple of months I have seen fresh road kill. But the prime indicator is if you have a dead sheep or a number of the sheep, how long it takes for that to be cleaned up can be directly related to the devil numbers.'

'Wombats have increased. We never saw any and now they've definitely increased here too. There's a big drop off in devils. Quolls seem pretty much the same. A lot of our quolls end up at the wildlife park … .'

'Rosellas are up. Devils are up. Snakes are down. I used to see five to six a day now it's five to six a year.'

Some animals were noted as increasing on some properties, but decreasing on others. For example, two graziers had observed a recent decrease in possum numbers, while six had observed an increase (see Appendix 2). The cause of local possum decline seemed to have been control measures:

'Possums, we get occasional permits to shoot them. A lot died and they haven't built up to the huge numbers they were a couple of years ago.'

One grazier suggested the possibility of future density-dependent disease mortality in possums:

'I saw 90 possums in one tree. Possums are heading the same way as the devils.'

Native hen populations were observed to have been highly volatile:

'Native hens are making a slight comeback – there used to be lots and lots of them.'

'There were thousands of native hens when I was a kid. A disease caused the population to crash overnight.'

'Native hens have increased. They dropped right down in the '80s and now they're up.'

'Native hens have increased. They decreased during the '90s. As a kid there were millions of them.'

The nature conservation implications of an observed increase in the numbers of cats concerned several graziers:

'I think cats are the cause of reduction in quolls etc. I shoot every one I see. One night he (the worker) shot seven cats, some had kittens. They're deadly on birds and little things like bandicoots and native hens. We used to have a lot of native hens.'

> 'Cats, there are hundreds of them, they've increased. They're a problem for native wildlife. They wiped out the bandicoots. We used to see them (bandicoots) on the lawn at night-time.'

> 'Now we have no bandicoots and ground nesting birds. The cats are mainly to blame. People don't realise the number of cats around.'

> 'There's been a decrease in green parrots and wattlebirds. We got rid of our domestic cats because we lost the birds in the garden. We've lost about 80 per cent of the birds since I was a kid from cats.'

As is indicated by the above quotes, most graziers had a sympathetic attitude towards native animals that were not a major economic burden. The threatened Tasmanian subspecies of the wedge-tailed eagle was the subject of several such comments:

> 'We had a lot of eagles until they were electrocuted. I saw two yesterday but there used to be more than that. I only know one farmer who shot eagles. Everyone else is proud of them and tries to protect them.'

> 'A wedge-tailed eagle took a small dog – dropped it as it couldn't lift off. I don't begrudge the eagles a lamb or two.'

> 'We've got more eagles now. It's good to see the eagles. You wouldn't see one 10 years ago.'

> 'Next door leaves dead trees and hollows for nesting sites and wedge-tailed eagle nests.'

Devils, quolls, bettongs and eastern-barred bandicoots have all been observed by graziers (see Appendix 2) as species that occur on the runs and all are of conservation concern. Several graziers mentioned the threatened ptunarra brown butterfly (alias brown tussock moth), which utilises *Poa* tussocks:

> 'We've got the brown tussock moth. The population has increased since fencing the bush block off in 1983. We went out shooting possums the year before last and there were lots of them (moths).'

In the 1998 ABS survey, 17 per cent of respondents reported knowing of rare or threatened species on their properties. The species that were mentioned included the green and gold frog, quoll, eastern barred bandicoot, wedge-tailed eagle and ptunarra brown butterfly (*Oreixenica ptunarra*).

Nature conservation management

Much of the management directed towards the maintenance or improvement of wool production from the runs also benefits the conservation of native species and communities. The culling of exotic animals and the control of exotic weeds are overwhelmingly positive for nature conservation. As will be further explored in Chapter 5, the impacts of grazing and fire regimes are highly contingent. The addition of inputs, such as fertilisers, is generally negative. The actions discussed below are those directed to nature conservation ends.

The 1998 ABS survey asked respondents what, if anything, they were doing about their dieback problems. Out of the 140 respondents who had tree decline, 131 had undertaken at least one action to remedy the problem. Of these 131 landowners, 52 per cent had culled possum and rabbit populations, 20 per cent had controlled patterns of grazing, 19 per cent had fenced areas of native bush, 16 per cent had revegetated with native trees, and 14 per cent had planted introduced trees. Only five of the 131 reported that their efforts were very successful. These five all reported culling possums and/or rabbits, and one noted that fencing of native bush was particularly successful in overcoming tree decline. Comments from this group included:

> 'Tree decline was noticeably reduced since the culling of possums'

and

> 'Fencing was very successful, but there was limited success with revegetation.'

Another 37 respondents found the response to their actions adequate, while 45 thought that their action/s had failed:

> 'We moved the kitchen in to here. We saw that two to three big trees had died and we decided that if we lived long enough we would be living in a desert and have no trees to look at. We fenced it off and they're still dying.'

Tree planting was a frequently mentioned conservation activity (see Appendix 2):

> 'My son has done a course. He strikes his own plants and grows them up in a possum-proof yard. He uses a deep rip tilling machine and planter. He planted 5000 trees a year for five years. After two years he got good survival but there were a couple of terrible patches that needed replanting. We tried to use electric fences but they were no good, couldn't keep the deer or possums out.'

Establishing trees is difficult (see Chapter 4):

> 'Possums and deer eat new trees. We tried planting seedlings in shelterbelts a few years ago, but they were all eaten by possums. Deer rub their antlers on new trees … .'

> 'Tree regeneration is something that worries me a lot. I don't know how we're going to overcome it, except by fencing areas off. Hand planting I have not found to be successful. I have probably planted in the vicinity of 2000 trees, and found maybe 200 to still be alive. If the drought doesn't

> get them, the deer will get them, if not the deer, the rabbits. I'm seriously unsettled about the denudation of the trees, but I can't see any way around it. The only way to do it is to fence off all areas, full stop.'

Bush regeneration, largely meaning tree regeneration, was the nature conservation management activity most mentioned by the 48 graziers (see Appendix 2). Half of the 'improved country set stockers', a third of the 'improved country rotators' and 'high and moist country graziers', and a fifth of the 'warm country graziers' said that they were regenerating a part, or parts, of their bush (refer to Table 3.1), largely by taking stock off it until regeneration was established:

> 'My ambition, if my son does all the hard work, is to fence off a lot more of the good quality areas to allow for longer recovery from grazing. I know if I'm overdoing it (grazing) if the sheep are in poor condition or are dying.'

There was a tendency for those who said that they were regenerating bush also to say that they were fencing off bush. Those graziers who said that they fenced bush also tended to say that they cropped and used the bush as a drought reserve. Properties with fenced bush had larger proportions of improved pasture than those without (43.48 per cent compared to 27.05 per cent). These relationships suggest that those who least rely on their runs for their income are the ones most likely to fence off parts of their bush for regeneration. One grazier elaborated on this phenomenon:

> 'Those landowners who have a higher proportional carrying capacity from improved pastures compared to their bush runs are more able to manage the runs according to the conservation needs of the bush. That is, they are able to spell their bush runs in spring and summer because there is plenty of surplus spring growth to utilise on the paddocks – this could be called the improved pasture subsidy (to the bush runs).'

> 'Increased cropping intensity on some properties can upset the pasture/bush ratio and result in either the bush becoming overgrazed during spring and summer or no grazing on the bush at all.'

The Tasmanian Regional Forest Agreement of 1997 set up a Private Forest Reserves Program (PFRP) designed to fill gaps in the conservation of forest communities and species on public land. This voluntary program enabled payments for the establishment of a permanent covenant on the title, and an associated management agreement, on areas of conservation-significant forest. An upfront payment, adjusted to the degree of conservation significance of the covenanted forest, and technical help, were provided. In 1998 a high proportion of respondents to the ABS survey (43 per cent) indicated that they were not willing to sign a legally binding voluntary agreement. Only 9 per cent of respondents indicated that they would be willing to sign such an agreement, the remainder being uncertain. However, if financial assistance for farm costs was part of such an agreement 56 per cent were favourably inclined to sign up, and only 19 per cent definitely would not have been involved. Less than 10 years later, 17 per cent of the

48 graziers stated that they had a covenant on part of their runs. The graziers with covenants were evenly spread among the enterprise groups with the highest proportion being among the 'warm country graziers', and the lowest among the 'improved country set stockers', a reverse pattern to fencing off (see Table 3.1).

Covenanting was not necessarily seen to involve a major economic sacrifice:

> 'I have a lot of areas of negligible grazing value, but the sites still have some value for them (PFRP) to offer me money for conservation. The runs hadn't been mined by grazing, and I won't feel much loss as it was non-productive bordering on a lot of rough bush country.'

> 'The covenanted land is not grazed. It's not that good for grazing. It has a low stocking rate, but it's of value for shrubs and trees such as peppermint and white gum.'

> 'I let the stock in, in winter, if I want to, in my covenanted block, as part of the agreement. Once or twice I put stock in there for shelter. I got an allocation to spray the gorse and I paid the balance out of my own pocket and got no (eucalypt) regeneration. Clearing it will remove the gorse.'

> 'It's quite nice to have an area to "have an area" and it wouldn't run a lot of stock anyway.'

However, sacrifice there was:

> 'We put ourselves out a little bit for it (the covenant). We provided the labour. We shut the country up, and the council recognised this in dollars, not enough to make a big difference to balance sheet. It's a gesture that shows genuine appreciation about it. We have two PFRP reserves and two others forced upon us when we bought … land.'

Those who said that they covenanted also tended to remark that the bush was a source of 'ferals', having experienced wild animal population build-ups in their reserved land (see Chapter 2).

Of those who had covenants put on their titles by 2003 as part of the PFRP, not necessarily farmers, 54 per cent valued their permanence:

> 'They are on the title, hence should be secure against the transient whims of government.'

> 'In perpetuity, binding in law, example to others as to how they may also contribute to conservation.'

Twelve per cent valued the financial support:

> 'The only program I know of that gives financial assistance for forest protection.'

Ten per cent valued the exclusion of logging:

> 'Knowing that, when we are dead, this place will not be logged for woodchips.'

Six per cent valued the continuation of private ownership:

> 'The ability to protect sensitive country, yet leave it in responsible private ownership.'

Of those who withdrew from negotiations on a covenant, 35 per cent largely did so because they perceived that a covenant would lower their property values. This perception was not necessarily correct. One grazier family found that their bush block in the central highlands was worth much more as a lifestyle block than it could ever be worth for farming/forestry.

Sixty-three per cent of those who withdrew from negotiations saw greater financial support as a necessary improvement to the program:

> 'I had a lot of gorse around my block that the documents said I had to control from the reserve. I was given sole responsibility to remove it. In my opinion they expected too much. It was not even the obligations that were a concern, it was meeting them. All you need is someone who enforces the covenant on you and you could find it very expensive in time and money to do these things.'

Twenty per cent of those who discontinued were uncomfortable with the perpetual nature of the covenant:

> 'We are asked to tie up land for ever.'

Some just did not like the conditions that would have been imposed:

> 'Three hundred acres not grazed, a rough area fenced off, was going into PFRP but they didn't want hunting with dogs.'

Some graziers regretted their participation in PFRP:

> 'PFRP offered dollars. I took advantage; wouldn't do any more. I don't like their agreements. It's all run by national parks type people. Their ideas don't include farming.'

> 'Covenant good from wildflower point of view but more of a nuisance than it's worth.'

The PFRP and the earlier purchase and leaseback of the Tom Gibson Nature Reserve have been very successful in improving the security of many rare or threatened species and plant communities on the runs, while still falling well short of reservation targets, such as those developed for the Regional Forest Agreement (Kirkpatrick 1998). The shortfall relates to funding levels. By 2006, the PFRP has achieved a greater percentage of its area target than the percentage of funds granted over those requested to achieve it.

Conclusions

A high proportion of Tasmanian wool growers appear to have a conservation ethic, an ethic that is grounded in their experiences of working with land, stock and nature, and a particular political history:

> 'Tasmanian farmers are somewhat unique in that many (if not most) farm enterprises have areas of native forests, woodland or grassland integrated into their operations. The TFGA (Tasmanian Farmers and Graziers Association) is the only NFF (National Farmers Federation) affiliate to have a forestry commodity committee in its structure. It was TFGA in arguing for removal of the export woodchip quotas that eventually initiated the RFA process in Australia. It argued in the mid '80s that conservation outcomes could be better achieved if they were properly identified and paid for, rather than applying arbitrary export quotas. While the experience has not always been a positive one, the engagement resulted in many Tasmanian farmers becoming better informed about the scientific principles of bush conservation, ethical issues involved and the principles of duty of care.'

Graziers seem to perceive that existence is as much about death as about life, and that the achievement of particular conservation goals is not without cost to themselves and some of the rest of nature. Above all, they are practical in their conservation-related actions, tending to try to avoid negative impacts on their enterprises from their conservation activities. Their comments have indicated some of the manifold complexities and contingencies of the relationships between people, sheep and nature. The following chapters further explore these complexities and contingencies in the context of trees on the run, remnant conservation and the effects of sheep grazing on the conservation of native ecosystems.

Chapter 4

Trees on the run

J.B. Kirkpatrick, D. Wilson, A. Meiss, A. Mollon and K.L. Bridle

Summary

Tasmanian tree dieback seems largely caused by prolonged drought, although agricultural activity, noisy miners and possums cannot be totally absolved from blame. Tasmanian woody plant thickening has been attributed to either the cessation of burning or the cessation of grazing. In a study area in the northern Midlands tree cover is shown to have increased between 1968 and 2002 where it was initially sparse, and to have decreased where it was initially dense. This pattern is attributed to the combined effects of less frequent burning and prolonged drought in the period, compared to earlier decades. Increases in woody plant cover can lead to changes in woody plant composition, with eucalypt dieback more likely where wattles, she-oaks and prickly box thicken. Although cultivation before planting or direct seeding helps in tree establishment, ripping and breaking up of the soil column have the opposite effect. There is evidence showing that successful planting or seeding of trees has been aided by the use of any two of post-planting chemical weed control, mulching and more than one watering. The use of stakes and guards in plantings appears irrelevant to their success.

Introduction

Tree dieback, tree thickening and successful tree planting are matters of considerable concern for most of the wool-growing community of Tasmania (see Chapter 3). This chapter reports scientific investigations of changes in tree cover, and the factors influencing success in tree planting, in the drier parts of Tasmania.

Causes of change in tree density

There has been a disturbing dieback of trees in the sheep grazing country of subhumid temperate Australia (Landsberg and Wylie 1983; Reid and Landsberg 2000; Close and

Davidson 2004). The rate of loss among isolated and scattered trees in rural Australia has been estimated at 0.54–2.5 per cent per annum (Freudenberger and Ozolins 2000), suggesting that their total loss could feasibly occur in 40–185 years. This is partly because trees retained during extensive clearing that occurred through the 1800s are reaching their maximum lifespan of 400–550 years. Paddock trees are also thought to be subject to a high rate of mortality due to the interaction between elevated nutrient loads and insect attack, salinity and soil compaction (Gibbons and Boak 2002). Eucalypts generally do not regenerate where ground is grazed heavily or cultivated (Bennett *et al.* 1994), so paddock trees in these areas are not replaced by seedlings. However, dieback has not been confined to isolated paddock trees in seas of crops or improved pasture. It has also been widespread in vegetation remnants and in runs. Tree dieback in Tasmania is widespread and variously explained (Grice 1995; Neyland 1996; Kirkpatrick *at al.* 2000; Close and Davidson 2004; Chapter 3).

There is strong circumstantial evidence that drought plays a major role in tree dieback in Tasmania (Kirkpatrick and Marks 1985; Neyland 1996; Kirkpatrick *et al.* 2000; Close and Davidson 2004; Chapter 3). Summed up, it is that the period since 1978 has been the driest on record in eastern Tasmania (see Figure 1.11), particularly in late summer/early autumn (Kirkpatrick *et al.* 2000), and the incidence of dieback is concentrated in the driest parts of the state (Neyland 1996) and on shallow soils over massive bedrock (Kirkpatrick and Marks 1985). Nevertheless, watering of trees had no effect on dieback in experimental conditions at three sites in the Midlands, an outcome posited to be the result of previous adjustment to the long drought through stand thinning and the small amount of extra water made available to the trees (Kirkpatrick *et al.* 2000). The dieback of cider gum in run country on the eastern Central Plateau (Potts *et al.* 2001) has occurred on lower valley slopes in areas with mean annual rainfall exceeding 900 mm, and may be more a response to the effects of increasing daily minimum temperatures on invertebrate predation (Kirkpatrick and Gibson 1999) than a reflection of drought.

There is also some circumstantial evidence that eucalypt dieback can be caused by noisy miners (*Manorina melanocephala*) wherever they establish breeding colonies. The presence of noisy miner colonies in vegetation remnants in subhumid lowland Tasmania has been shown to be associated with tree dieback, an absence of a dense understorey and an absence of smaller birds, such as honeyeaters and robins (MacDonald 2001; MacDonald and Kirkpatrick 2003). The same relationships have been observed elsewhere in eastern Australia (Bennett and Ford 1997; Grey *et al.* 1998). Scientists in Victoria removed noisy miners from a remnant and subsequently recorded the invasion of small native birds (Grey *et al.* 1997; Clarke and Schedvin 1999). These small birds have been shown to be able to coexist with noisy miners where there is a dense understorey. However, if the vegetation consists only of trees and a grassy ground layer, there is nowhere for them to shelter from the aggression of noisy miners out to protect their food resources (Dow 1977). Like many of the smaller birds, the noisy miners feed on invertebrates that in turn feed on eucalypts. It has been suggested, without definitive proof, despite experimental attempts (Clarke and Schedvin 1999), that they do not eat sufficient of the invertebrates to prevent defoliation of the trees in their colonies, thereby causing dieback. The alternative hypothesis is that the conditions that motivate noisy miners to establish colonies are also the conditions that encourage dieback.

Brushtail possums also tend to favour the edges of forests and remnants, especially edges adjacent to improved land and cropland. There is experimental evidence that possums play a role in eucalypt dieback (Kirkpatrick *et al.* 2000). More specifically, there is evidence that keeping possums out of trees can result in either tree recovery, or a lesser rate of deterioration than occurs in trees to which possums have access, at two out of three experimental dieback sites in the Midlands. At the third site possum exclusion had no effect, the dieback being attributed to self-thinning (Kirkpatrick *et al.* 2000).

Woody plant thickening has also been recently observed in grassy country in Australia and elsewhere, with theories to explain it including increased carbon dioxide in the atmosphere, climatic changes and fire and grazing regime changes (Bennett 1994; Archer 1995; Brown and Carter 1998; Bowman *et al.* 2001; Sharp and Whittaker 2003; Kirkpatrick 2004). On the Tasmanian runs, graziers noted woody plant thickening only half as much as they did tree dieback (see Chapter 3).

Tasmanian graziers often attribute increases in eucalypts to a decreased incidence of fire. For example Jim McEwen of 'Trefusis' observed:

> 'We haven't burnt anything since 1967. The laws have changed; liability is a much bigger issue. Lack of fire has changed the composition of the bush.' (Gilfedder *et al.* 2003, p. 8)

An alternative explanation that has been advanced for tree thickening in Tasmania is a lack of grazing. The woody plants of the Domain, a grassland and grassy woodland remnant in the city of Hobart that has been totally ungrazed since the mid twentieth century, has thickened remarkably since grazing ceased, with the thickening being most extreme where the vegetation was most burnt (Kirkpatrick 2004). It seems likely that both fire, and grazing after fire, may be necessary to prevent woody plant thickening in these particular vegetation types.

Tree thickening and dieback

A study area, consisting of the bush runs on three large properties to the south-east of Ross in the northern Midlands (see Figure 4.1), was selected to observe changes in tree cover. It was selected because of the different history of burning and grazing regimes on each property, the diversity of topography, geology and vegetation types, and the presence of altitudinal (300–600 m), and, therefore, climatic variation. The idea was to compare tree cover before the dry period that started in the late 1970s to tree cover in the twenty-first century and to see which environmental and management variables

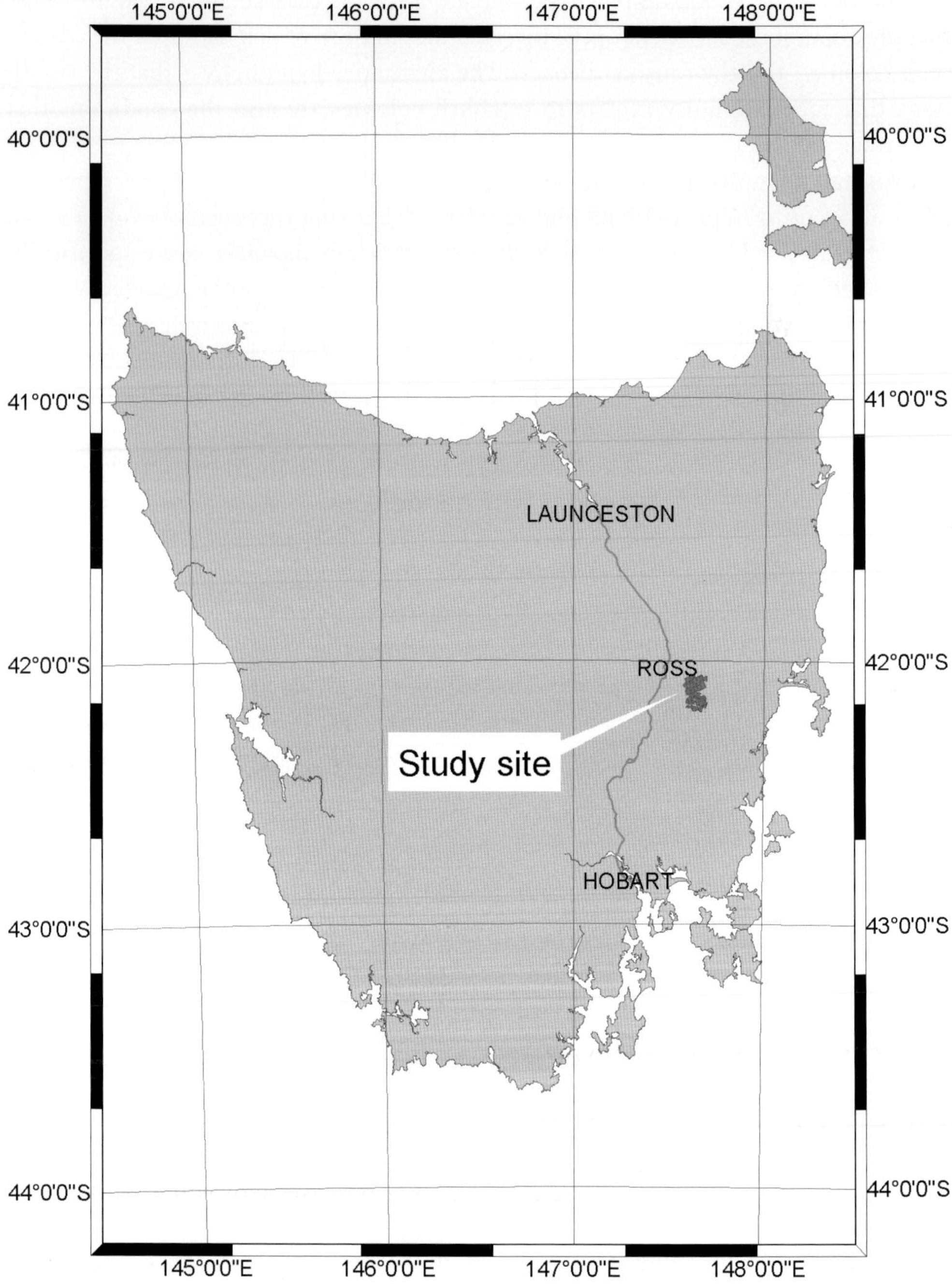

Figure 4.1. Study site for monitoring of changes in tree cover.

were most related to any change. Aerial photograph analyses, and modelling of the correlates of change, were used for this purpose.[1]

The best general linear model to explain 1968 canopy cover consisted of property,[2] elevation,[3] winter radiation,[4] geological class[5] and aspect.[6] The results for each variable

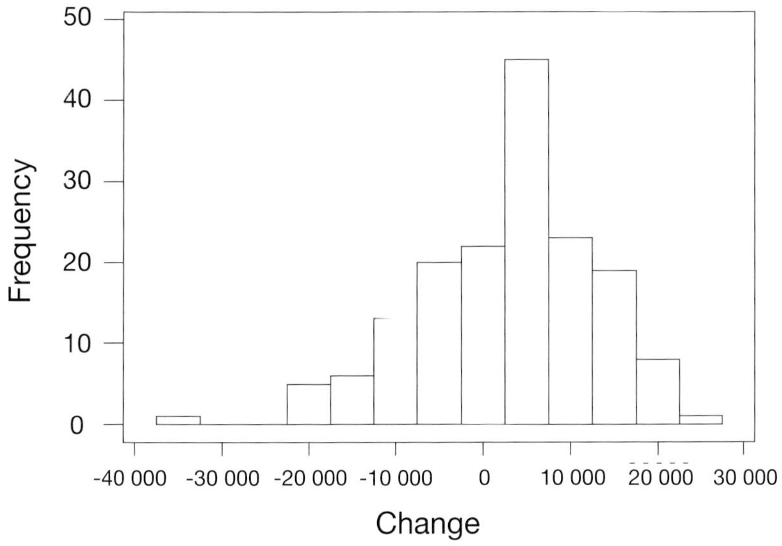

Figure 4.2. The frequency distribution of canopy cover change (m^2/6 ha).

effectively hold the other variables constant, meaning that all else is equal. The strong influence of property in the model probably reflects differences in their long-term management regimes. These had resulted in greater tree canopy cover on one of the properties in any particular environment. The 1968 tree canopy cover also strongly related to elevation, which probably reflects the positive effect of altitude on precipitation. Winter radiation is negatively related to canopy cover, possibly reflecting differences in soil moisture, related to higher moisture loss through evapotranspiration on the north-facing slopes, which receive more radiation than the south-facing slopes. Quaternary sediments had lower canopy cover than dolerite, which in turn had lower canopy cover than sedimentary rocks. The Quaternary sediments were in the valley bottoms, which naturally had the most open vegetation (see Chapter 1). Dolerite and sedimentary rocks were on the slopes. Dolerite soils hold less available moisture during the dry season than soils on sedimentary rocks (Kirkpatrick and Marks 1985), so a lesser tree cover could be expected. The more north-westerly aspects had lower canopy cover than the more south-easterly aspects. North-westerly slopes receive most sun in the afternoon and are most fully exposed to drying north-westerly winds (Kirkpatrick and Nunez 1980).

The sample showed an overall increase in canopy cover, by 8 per cent between 1968 and 2002, with considerable variability, which extended from marked decrease to marked increase. The frequency distribution of change was slightly negatively skewed (see Figure 4.2). There was a high degree of spatial contagion, with canopy decrease showing a tendency to be concentrated in the east of the study area (see Figure 4.3).

The best model of change consisted of 1968 canopy cover,[7] vegetation type,[8] slope[9] and geology.[10] Those areas with high tree canopy cover in 1968 tended to lose cover by 2002, while those with low cover in 1968 tended to have more cover in 2002 (see

Figure 4.3. The spatial pattern of canopy cover change (m^2/6 ha).

Figure 4.4). This result, and the tendency of tree canopies to increase more on steeper slopes than gentler slopes and more on dolerite than on other substrates, can be most parsimoniously interpreted as a product of the combination of lower rainfall and less frequent fire in the period 1968–2002 than in earlier times. A marked reduction in available soil moisture is likely to be reflected in a reduction of tree cover, there being a well-known relationship between vegetation biomass and moisture up to the point where water is not limiting. Where trees were closest to equilibrium with moisture availability in 1968, the canopy cover was likely to be higher than in areas where tree cover had been reduced by a long history of burning in combination with grazing.

Sheep grazing is concentrated on the gentler, drier north-facing slopes in the study area (Leonard and Kirkpatrick 2004), where fire is more likely to carry than

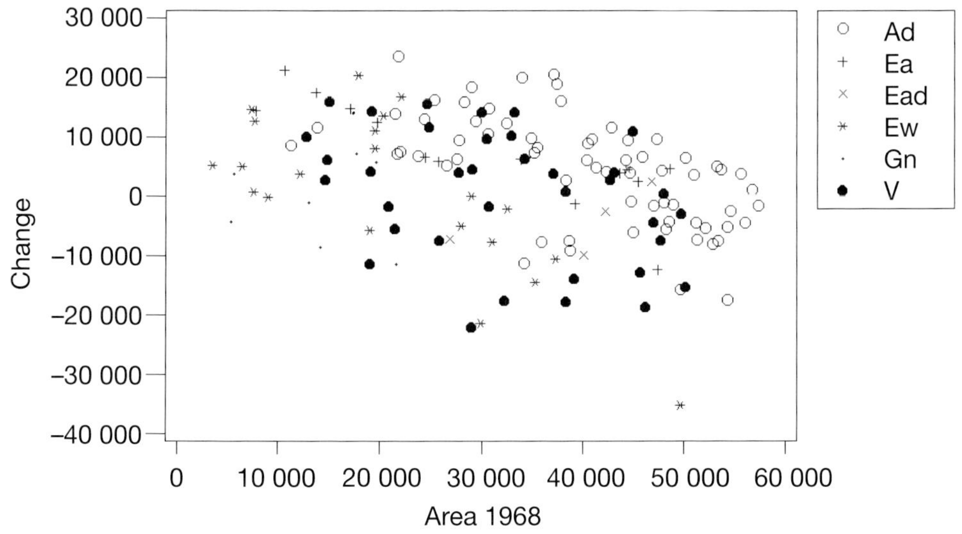

Figure 4.4. The relationship between canopy cover change (m^2/6 ha) and the canopy cover area in 1968 (m^2/6 ha), showing vegetation types. Ad = *Eucalyptus amygdalina* forest on dolerite, Ea = *E. amygdalina* woodland, Ead = *E. amygdalina* woodland on dolerite, Ew = *Eucalyptus viminalis* woodland, Gn = grassland, V = *Eucalyptus viminalis* grassy forest.

on the steeper, moister south-facing slopes. Thus tree cover was more likely to be in equilibrium with moisture availability on the steeper slopes, meaning that a decrease in tree cover due to a decline in rainfall was more likely to occur there, as is intimated by the model. The north-facing slopes on dolerite have grassy vegetation that forms continuous fuel, and can be readily burned at a high frequency. The vegetation on the north-facing slopes on sedimentary rocks does not accumulate fine fuel as rapidly, has a naturally denser tree layer because of the higher levels of moisture available in the sandier and deeper soils (Hogg and Kirkpatrick 1974; Kirkpatrick and Marks 1985), and is less productive in terms of sheep fodder than the vegetation on north-facing slopes on dolerite. All these factors would combine to make it more likely that tree cover would be dense enough on sediments to suffer decline with a decrease in rainfall.

It is unlikely that changes in grazing regimes have caused the increases in tree cover, despite the well-known potential for them to do so (Bennett 1994; Brown and Carter 1998; Costella *et al.* 2000; Sharp and Whittaker 2003), as discussions with the owners indicate that the properties have been grazed in the same way for many decades. However, the known reduction in fire frequency is likely to have allowed suppressed eucalypt seedlings to escape the grazing zone, and have encouraged the growth of wattles, which can establish new individuals from root sprouts. This thickening would not be limited by drought where trees had been reduced in cover by past burning and grazing regimes. Therefore, the analyses lead to some of the same conclusions made by graziers on the causes of dieback and thickening (see Chapter 3).

Vegetation dominated by white gum tended to dieback more, and thicken less, than vegetation dominated by black peppermint (see Figure 4.4). This difference between species is consistent with their known susceptibilities to dieback (Neyland 1996; Chapter 3) and their distributions in relation to water availability (Fensham 1989). In the study area black peppermint forest on dolerite occurred at higher elevations[11] and on steeper slopes[12] than white gum forest.

This means that both thickening and thinning of canopies can occur simultaneously within the one forest, related to variation in all of initial state, dominant species and environmental variables. Thickening may be reversible if fire is reintroduced as a management tool in these forests and woodlands, and is followed by grazing. However, if thinning is largely the result of climatic change, there is little to be done but accept it.

Compositional change in the thickening process

The graziers who talked about increases in canopy cover in their run country tended to mention increases in wattle and prickly box rather than an increase in eucalypts. On the Domain, in Hobart, where marked increases in tree density and cover occurred in the late twentieth century (Kirkpatrick 2004), increasing dieback of white gums was observed (see Figure 4.5). This appeared to be associated with the areas where woody plants had most thickened. To test the validity of this observation, nine sites varying in environment and the level of competition for resources were subjectively selected (see Figure 4.6). Within each site, data on the health of trees and the environment in which they occurred were collected.[13]

Mean eucalypt health scores for the nine areas were 80 per cent explained by the positive effects of total basal area and mean individual basal area in the best model.[14] Therefore the major influences on tree health were competition from other trees for moisture and maturity, with large trees and stand openness associated with good health. The same relationship was apparent for white gum alone. The lack of influence of environmental variables related to soil moisture, which is likely to vary enormously on the Domain in response to variations in evapotranspiration related to slope and aspect (Kirkpatrick and Nunez 1980), suggests an extreme water stress event. This indeed was the case. In 103 years of record keeping, 1999 and 2000 were the eighth and tenth driest years respectively. None of the other 10 driest years occurred consecutively. The tendency for large trees to be healthier once competition was held constant may be related to their suppression of the ground layer and their likely occupation of particularly favourable microhabitats within an area.

The implications of these observations and analyses for the run country are that the impacts of increases in canopy cover on fodder availability on the ground layer are likely to be greater than the magnitude of the change, as, under drought stress, tree species that suppress grass growth, like she-oak, or interfere with grazing, like prickly box, are likely to replace the open-canopied eucalypts.

Figure 4.5. Dieback of white gum (*Eucalyptus viminalis*) and thickening of she-oak (*Allocasuarina verticillata*) on the Domain, Hobart (Jamie Kirkpatrick).

Figure 4.6. Sites on the Domain selected for a study of tree health related to environment.

Table 4.1. Influence of environment and treatment on tree establishment (% of planted area stocked with trees)

Variable	Mean % area stocked (n)		P
	Yes	No	
Soil column broken up	59 (27)	**82** (15)	0.010
Mean annual precipitation > 600 mm	**74** (37)	51 (11)	0.018
Mean daily max. temp. warmest month < 24°C	**73** (35)	51 (11)	0.024
Soil ripping before planting	60 (25)	**78** (23)	0.026
Watered more than once after planting	**79** (21)	61 (27)	0.029
Mean daily min. temp. coldest month < 0.4°C	59 (24)	**78** (22)	0.034
Chemical weed control after establishment	**83** (12)	64 (36)	0.044
Mulch used	**78** (22)	61 (26)	0.045
Mechanical weed control after establishment	**81** (13)	64 (35)	0.072
Ground pre-prepared for planting	**71** (43)	47 (5)	0.080
West, north-west or north aspect	**83** (9)	65 (39)	0.086
Slope < 1°	64 (33)	**78** (15)	0.125
Insects a problem	**73** (19)	61 (29)	0.165
Pest control undertaken	63 (24)	**74** (24)	0.211
Ground prepared > 2 months before planting	**75** (25)	66 (18)	0.263
No mechanical preparation for planting	**78** (10)	66 (38)	0.272
pH > 6.0	61 (12)	**71** (36)	0.279
Plants watered in	**73** (18)	63 (24)	0.285
Combined mechanical and chemical preparation	**74** (28)	65 (20)	0.287
Seedlings staked	61 (11)	**71** (37)	0.328
Native trees and/or understorey in planted area	59 (8)	**70** (40)	0.330
Cultivation before planting	**76** (10)	67 (38)	0.361
Residual herbicides used before planting	**76** (7)	67 (41)	0.437
Orientation to prevailing wind direction 90°	66 (29)	**72** (19)	0.453
Seedlings planted (not direct-seeded)	67 (42)	**76** (6)	0.511
Fertiliser used	**72** (7)	68 (41)	0.744
Sheltered from north-west winds	**72** (5)	68 (43)	0.778
Knockdown herbicides used before planting	**69** (34)	67 (14)	0.800
Tree guards used	68 (37)	**70** (11)	0.848

The P values are from ANOVA. Bold figures denote the higher success rate.

Putting trees back

A common response to the dieback of paddock trees has been attempts to establish new trees by either the planting of seedlings or direct seeding (see Chapter 3). While there have been many guides to the successful establishment of trees on farms (for example Oates and Clark 1987; Venning 1988; Cremer 1990; Heinjus 1992; Kirkpatrick and Gilfedder 1999), there have been few studies that have assessed the utility of

Table 4.2. The mean percentage of planted area stocked with trees for the combinations of significant post-planting treatments

Treatment/s	% stocked (n)
Chemical weed control and mulch	90 (1)
Chemical weed control and > 1 watering	90 (1)
Mulch and > 1 watering	83 (8)
Chemical weed control	81 (4)
Chemical weed control and > 1 watering and mulch	77 (7)
> 1 watering	72 (5)
Mulch	68 (6)
None of the three treatments	51 (16)

different techniques in achieving tree establishment in the drier parts of Tasmania. Pinkard (1992) assessed direct seeding techniques, but unfortunately experienced an extremely dry period for her experiments. Much of the advice available from published sources is conflicting, so, in an attempt to see what really works, 48 farm plantings of more than 100 trees more than one year old, distributed throughout the drier parts of Tasmania, were assessed in 2001.[15]

Several environmental variables significantly influenced success (see Table 4.1). Success was greater where mean annual precipitation exceeded 600 mm, where the mean daily maximum temperature of the warmest month was less than 24°C and where the mean daily minimum temperature of the coldest month was greater than 0.4°C. These results are unsurprising, in that moisture availability, frost and heat stress are well-known causes of mortality in seedlings. There was no significant effect of the degree of naturalness of the vegetation on the site used for planting.

Most farmers prepared the ground for planting before the planting event, with the few who did not having less success than those who did, although the difference was not significant (refer to Table 4.1). However, ripping or breaking up of the soil column were related to significantly less success than on sites where such extreme soil disturbance did not take place (see Table 4.1). Ripping and breaking up of the soil column are often used to help establishment on more difficult sites, such as those with heavy clays or hard pans. It is not possible from the data to know exactly to what degree site conditions, or the act of disturbance, are responsible for lower success, although neither disturbance of the soil column nor ripping were significantly associated with the gentlest slopes, where most such difficulties are encountered. On average, the sites where less severe cultivation and/or chemical preparation was/were undertaken there were greater success rates than on those that were ripped or not prepared (see Table 4.1). The length of time between site preparation and planting or seeding was not significantly related to success.

Although direct-seeded sites had a slightly higher success rate than planted sites, the difference was not significant. Neither the orientation nor shelter of the plantings was significantly related to success (see Table 4.1). Fertiliser application made no difference. Neither did staking of seedlings nor the use of tree guards. In fact, both had a negative effect on average, suggesting a means to save on the costs of tree establishment. A once-off watering in of plants had no significant effect on success.

Treatment after planting or seeding was critical for success. Sites that were watered more than once after planting had significantly higher success rates than those that were not so watered (see Table 4.1). Chemical weed control significantly improved success, as did mulching (see Table 4.1). The difference in mean success between the most successful combinations of these treatments (chemical weed control and watering, chemical weed control and mulching) and none of these treatments was large (refer to Table 4.2). All three treatments in combination give good success but less than the most successful combinations (see Table 4.2), suggesting another means to reduce the costs of planting.

Although engagement in pest control did not have a significant effect on success (see Table 4.1), it is likely to have reduced losses that would otherwise have occurred in areas where pest animals were numerous. Plantings that were reported by farmers to have had problems with insects did not significantly differ in their success rates from those that had no problems, presumably because of a pesticide response, and possibly because fast-growing plants attract invertebrates.

The recommendations for cost-effective tree planting that logically flow from the above analyses are: cultivate – but do not rip – the site before planting; use any two of chemical weed control, mulching and more than one watering after planting; do not use stakes and tree guards.

Notes

1 Complete aerial photograph runs were available for 1968 (1:13 000) and 2002 (1:42 000). Two hundred and twenty-five six-hectare squares were randomly selected from the study area, with each paddock having a number of squares proportional to its share of the total area. Because shrubs, largely wattles, and eucalypts reflect less light than the grassy understorey it was possible to use digital image processing techniques (Kadmon and Harari-Kremer 1999) on rectified photographs, combined with careful manual checking of the relationship of grey levels in the images to tree cover visible on the images, to determine the cover of trees in each of these squares at each of the two times. The 1968 cover was subtracted from the 2002 cover to give a change value. Vegetation type, aspect (adjusted to a score of 1–5 from the dry north-west to the wetter south-east), altitude, winter radiation (using the formula of Nunez 1983), property and geological type for each square were obtained, or calculated, from geographic information system databases. More details of these measurements can be found in Wilson (2004). To work out the conjoint influences of many factors on 1968 canopy cover and change in canopy cover from 1968 to 2002, general linear and multiple regression models were created, using: 1) 1968 cover as the dependent variable and various combinations of geology type, aspect type, property, slope, winter radiation and elevation as the independent variables; 2) canopy cover change as the dependent variable and various combinations of geology type, vegetation type, aspect type, property, slope, winter radiation, tree canopy area 1968 and elevation as the independent variables. In general linear models, continuous variables were treated as covariates. The distribution of the residuals was examined to ensure satisfaction of the assumption of normality of residuals.

2 $F = 13.40$, $P < 0.001$.

3 $F = 12.98$, $P < 0.001$.

4 $F = 9.46$, $P = 0.002$.

5 $F = 6.94$, $P < 0.001$.
6 $F = 4.43$, $P = 0.005$.
7 $F = 57.01$, $P < 0.001$.
8 $F = 7.78$, $P < 0.001$.
9 $F = 8.77$, $P = 0.004$.
10 $F = 2.26$, $P = 0.048$.
11 ANOVA, $F = 5.69$, $P < 0.001$.
12 ANOVA, $F = 6.3$, $P < 0.001$.
13 The following data were recorded, in 2001, for all eucalypt trees: species, girth at breast height, position and health. The following scale was used to rate health: 1, dead; 2, dead with dead foliage; 3, epicormic shoots confined to main trunk; 4, epicormic shoots on branches; 5, mainly adult foliage with dead branches; 6, trees with normal adult foliage. Depending on the density of eucalypts, between 23 and 34 trees were measured at each site. Values of girth at breast height for each tree were converted to basal area and used as a measure of tree size. A mean tree health score was calculated for each of the nine areas. The mean basal area per tree for each species was also calculated. Data on the basal area of the vegetation as a whole were means of Bitterlich wedge estimates from randomly located quadrats within each of the nine sites. Environmental variables were the means for these quadrats within the tree health sampling areas. The same modelling process that was described in note 1 was used with the tree health score as the dependent variable and summer radiation, equinoctial radiation, winter radiation, slope, altitude, aspect (converted to the scale described in note 1), years since the last fire, the longest interval between fires, the number of fires 1965–2000, total basal area per unit area, mean individual tree basal area and mowed/not mowed as independent variables. See Figure 4.6 for the location of sampling areas for dieback.
14 $F = 16.1$, $P = 0.004$.
15 The key variable in planting success is the proportion of the planted area that is adequately stocked. This was estimated as a percentage for each of the 48 planting areas. The farmers provided details on their preparation, establishment and post-establishment management techniques. Most environmental data were obtained from published maps. The pH of the top 4 cm of the soil was measured with a CSIRO soil-testing kit. Because of difficulties with the normality of continuous independent variables, the relationships of these to the percentage of area stocked were visually examined to look for breaks in responses, and these breaks in responses were used to convert these variables into the yes/no form. For all independent variables one-way analysis of variance (ANOVA) was used to test whether the variation in success between each of the two classes significantly exceeded the variation within the two classes. For these analyses a chance of one in 20 or less ($P = 0.05$ or less) that the difference is not real is regarded as being significant. Exact probabilities are given in Table 4.1 in case the reader wishes to adopt a different level.

Chapter 5

Sheep and nature on the run country

J.B. Kirkpatrick

'Domestic grazing animals have been described as having a slasher out the front, four little cultivators underneath and a fertiliser spreader out the back (T.A. Wright, personal communication, 1998). When the stocking rate is low the slasher becomes fairly selective ..., the cultivators have little effect in the short term, and the fertiliser is unevenly spread around the paddock.' (Whalley 2005, p. 3)

Summary

Sheep grazing regimes, and their interactions with burning regimes, have led to local extinctions in the runs, but most native plant and animal species, including many that are threatened, coexist with sheep in unimproved country. Some plants even depend on sheep to reduce competition in situations in which it is not possible to allow native herbivores to build up their numbers. Different management regimes in the run country favour different sets of plant and animal species suggesting that there is no one right way to manage runs for nature conservation. This implies a need for spatial heterogeneity in run management at the landscape scale. In contrast, temporal heterogeneity is undesirable, as species adapted to the outcomes of one management regime may perform poorly in another.

Introduction

The introduction and establishment of a new species of herbivore usually results in changes to the ecosystem into which it is inserted. These consequences vary from the dramatic, as with the introduction of goats to remote oceanic islands (Coblentz 1978), to the apparently negligible, as with the human-assisted return of the horse to continental North America. The severity of the consequences of introductions has tended to be greatest where there are no native vertebrate herbivores that eat in the same vegetation layers, or where the introduced herbivore has dramatically different feeding patterns, and other environmental impacts, from those of any native vertebrate herbivore. If the ecosystem receiving the introduction is taxonomically and functionally similar to that in which the herbivore evolved, the ecological consequences are likely to be weaker than when origin and destination are dramatically different.

Australia is biotically the most different of continents. The hard hooves of placental herbivores were absent before the European invasion, the large herbivore niche being filled by marsupials with soft feet. Few of the species that formed native forage for the introduced sheep, cattle, horses and goats occurred elsewhere in the world, the widespread kangaroo grass, *Themeda triandra*, being a notable exception in its natural domination of southern African grasslands. The accelerated erosion associated with the waves of stock emanating from Sydney, Hobart and Port Dalrymple in the early nineteenth century (see Chapter 1) is consistent with a sudden and severe depletion in vegetation cover induced by European domestic animals. A severe reduction in the capacity of soils to infiltrate moisture, because of compression caused by stock, may also have contributed to this pulse of erosion. A subsequent decline in erosion rates is consistent with a recovery in vegetation cover, as more resistant native and exotic species filled the vacant spaces (Moore 1970). It is also consistent with a depletion of the sediment available for transport.

This chapter is not devoted to the dramatic beginnings of the interactions of sheep with nature in Tasmania (for early history of Tasmanian sheep, see Chapter 1), but rather the outcomes from the interactions in the early years of the twenty-first century. If nature is to be conserved in a landscape largely devoted to wool production, these outcomes must be understood. Given that there is a wide variety of ways in which the sheep, and the runs in which they graze, can be managed (see Chapter 2), the key question is: *Which management regimes benefit or reduce which native species and communities?*

Sheep and plant species

Hobbs and Huenneke (1992) have pointed out the species-specific nature of grazing impacts on plant species, citing the example of the study of Elkington (1981) on an English chalk grassland, which found that removal of sheep favoured several rare species, but caused the decline of another. These impacts take place directly and indirectly. Plant species vary in their attractiveness as food sources. The more palatable might be expected to be selected against in any pasture, all else being equal (Moretto and Distel 1999; Dorrough, Yen *et al.* 2004). Sheep tend to concentrate their grazing on the parts of any paddock that contain the most palatable plants, and work their way down. Commenting on mob stocking, one Tasmanian grazier said:

> 'On the first day the sheep eat the ice-cream, the second day the brussell sprouts.'

The indirect effects of sheep grazing on interference between plant species can be equally or more important. For example, the grazing of dominant tussock or shrub species often allows smaller plant species to persist (Wells 1969; Gilfedder and Kirkpatrick 1994a; Lunt 1997). In grazing-adapted systems, the exclusion of grazing animals is often a recipe for a dramatic reduction in species richness (Hobbs and Huenneke 1992). Sheep also disturb and compact the ground with their hooves, and, in set stocked paddocks, redistribute nutrients from the rest of the paddock to their camps (Greenwood and McKenzie 2001; Prober *et al.* 2002). They are effective agents of dispersal for many exotic and native plants. Seeds stick to their fleece or persist in their pellets.

Several Tasmanian native plants of fertile and dry country are now only found where sheep grazing is a very rare or absent event. They survive in cemeteries, farmyards, road reserves, city parks and domestic gardens (Kirkpatrick *et al.* 1988). The ecologically best known of these species is the soft peppercress (*Lepidium hyssopifolium*), a nationally endangered species that has only been observed once in natural vegetation in recent decades, under black wattle (*Acacia mearnsii*) trees in a seldom-grazed paddock in the Midlands of Tasmania (Pyrke 1994). Most of its surviving populations occur in vegetation with few other native species, in the bare zone under dense-canopied exotic trees on road reserves, in city parks and in farmyards, where it can occasionally be seen growing on colonial sandstone steps (Kirkpatrick and Gilfedder 1998). It, and the rare closely related annual herb, shade peppercress (*Lepidium pseudotasmanicum*), cannot survive sheep grazing or dense above-ground competition from other plants, but are drought-hardy, occurring where they are subject to intense root competition or where soils are non-existent or extremely shallow. They are prolific producers of seed that is stored in the soil, which enables persistence after occasional grazing events. Pyrke (1994) was able to resurrect a roadside population, which had lost its competition-suppressing exotic tree, by vigorous cultivation and removal of the exotic grasses that had colonised the bare zone.

Several perennial wildflowers are rarely found in Tasmania outside places protected from sheep grazing. These include lanky buttons (*Leptorhynchos elongatus*), showy copperwire-daisy (*Podolepis jaceoides*), alpine billybutton (*Craspedia coolaminica*) and spur velleia (*Velleia paradoxa*). Profusely flowering populations of alpine billybuttons, lanky buttons and showy copperwire-daisy occur in the vertebrate herbivore exclosures on Liaweenee Moor (see Figure 5.1). Seedlings establish outside the fence, but are usually removed by sheep grazing each summer, although a few hang on in the midst of prickly shrubs (Bridle and Kirkpatrick 2001).

The yam daisy (*Microseris lanceolata*), a staple in the Aboriginal diet in south-eastern Australia, was so common in grasslands at the time of the European invasion that its flowers turned the plains yellow (Gott 2005). Today it is most common where no stock graze, but nowhere is it as abundant as in the days of Aboriginal digging. Like most of the above species, it appears to require both freedom from competition and freedom from heavy grazing, especially sheep grazing, to occur in abundance.

In a meta-analysis of the response of plant species of the Tasmanian lowland runs to various levels of sheep grazing intensity, Kirkpatrick, Gilfedder *et al.* (2005) found

Figure 5.1. An alpine billybutton (*Craspedia coolaminica*) sheltering within the Liaweenee exclosure. The small flowers of scaly buttons (*Leptorhynchos squamatus*), a species that becomes more abundant with sheep grazing, can be seen outside the exclosure (Kerry Bridle).

that only five species had a negative response to sheep grazing in two or more data sets. Kangaroo grass (*Themeda triandra*) was one of these, a result consistent with the mainland meta-analysis of Vesk and Westoby (2001). Others were the common woodruff (*Asperula conferta*), the narrow leaf New Holland daisy (*Vittadinia muelleri*), the short-stem sedge (*Carex breviculmis*) and the introduced grass genus, *Bromus*. The data sets variously compared sheep grazed areas to areas with no sheep and more intensely grazed areas to less intensely grazed areas. Only the more abundant species at each site could be analysed statistically. The results indicate relative responses, rather than absolute responses. For example, a large part of Tasmanian native grassland has been mapped as wallaby grass-dominated grassland in which there appears to be little or no kangaroo grass. Yet, kangaroo grass often becomes dominant once sheep are excluded, indicating its continued presence in a severely shortened form (see Figure 5.2).

As on the mainland (Vesk and Westoby 2001), many species evidenced contradictory responses between sites, responding positively to more intensive sheep grazing on some sites, but negatively on others (Kirkpatrick, Gilfedder *et al.* 2005). These species included the Australian sheep-burr (*Acaena ovina*), cranesbills (*Geranium* spp.), hairy guineaflower (*Hibbertia hirsuta*), small St Johns wort (*Hypericum gramineum*), peachberry heath (*Lissanthe strigosa*), sagg (*Lomandra longifolia*), five-awn speargrass (*Pentapogon quadrifidus*), dwarf riceflower (*Pimelea humilis*), small quaking grass (*Briza minor*) and thistles (*Cirsium* spp.). The most likely reason for contradictory responses is that some species perform best at an intermediate level of grazing intensity. This means that at sites that compared the effects of no sheep grazing to moderate sheep grazing they would perform better under moderate sheep grazing, because of the reduction of more palatable competitive species, while at sites that compared moderate

Figure 5.2. Sheep-grazed lawn dominated by wallaby grass (*Austrodanthonia* spp.), with kangaroo grass (*Themeda triandra*) inside a small exclosure (Kerry Bridle).

sheep grazing with heavy sheep grazing they also performed better under moderate sheep grazing, because with heavy grazing they became the 'brussell sprouts'.

In the Tasmanian meta-analysis, five species of native plant (wallaby grass, Australian carrot (*Daucus glochidiatus*), scaly buttons (*Leptorhynchos squamatus*), small poranthera (*Poranthera microphylla*), native bluebell (*Wahlenbergia* spp.)), an introduced grass (*Vulpia* spp.) and clover (*Trifolium* spp.) were consistently more abundant in the more heavily grazed parts of sites (Kirkpatrick, Gilfedder *et al.* 2005). In most cases, the relative freedom from shading in the more heavily grazed areas is likely to be the cause of the consistent response, although the small poranthera and *Vulpia* are also reputed to be highly unpalatable, in contrast to clover and wallaby grass.

There are several species of greater rarity than those listed in the previous paragraph that appear to be largely dependent on heavy grazing by sheep for their survival in the run country. The ecologically best known of these species is the grassland paperdaisy (*Leucochrysum albicans*). This spectacularly flowering species (see Figure 5.3) is not found in the soil seed bank (Gilfedder and Kirkpatrick 1993b) but its plumed seeds can be dispersed long distances by wind (Gilfedder and Kirkpatrick 1994b). It disappears rapidly from a site if new bare ground is not available for colonisation every few years (Gilfedder and Kirkpatrick 1994a, 1994b). In pre-European times bare ground

Figure 5.3. A profusion of the white flowers of grassland paperdaisy (*Leucochrysum albicans*) in a heavily grazed paddock (Matt Appleby).

would have been dispersed throughout the landscape, produced by the digging of animals, fire, and wind and water erosion. While bare ground is more common now, it is almost entirely devoted to the growth of crops, which out-compete the paperdaisy. The distances between uncultivated sites with bare ground have increased, as most pastures are managed to maintain high levels of cover.

Gunn's mignonette (*Stackhousia gunnii*, see Figure 5.4) is another rare species that disappears when the grass sward overwhelms bare ground for several years (Gilfedder and Kirkpatrick 1998b), as do the grassland cupflower (*Colobanthus curtisiae*, Gilfedder and Kirkpatrick 1996) and the many rare orchids of the Midlands. None of the above species have been observed to persist in soil seed banks (Gilfedder and Kirkpatrick 1993b; Friend, Cameron *et al.* 1997). The period of persistence of the tubers and bulbs of geophytes, such as the orchids and Gunn's mignonette, is not known.

Several tiny annual daisies are particularly susceptible to local extinction where bare ground disappears. These include southern buttons (*Cotula australis*), moss sunray (*Hyalosperma demissum*), grass cushion (*Isoetopsis graminifolia*), soft bowlflower (*Millotia tenuifolia*) and dwarf sunray (*Triplodiscus pygmaeus*). In the Tom Gibson Nature Reserve several of these species are most common in scalds in the heavily grazed native grassland around the shearing shed (Fensham 1991).

Figure 5.4. Gunn's mignonette (*Stackhousia gunnii*) (Kerry Bridle).

Large populations of native herbivores can be as effective in creating bare ground as large populations of sheep, as has happened in Maria Island National Park, where forester kangaroos, wallabies and wombats ate down the sward until they starved to death on a diet of dirt. Wombats are particularly effective in creating bare ground, as they dig for roots and tubers and excavate impressive burrow systems. Bare ground can also be created by the diggings of echidnas, potoroos, brown bandicoots, eastern barred bandicoots and bettongs. Pyrke (1994) observed that, in a year, 2.9 per cent of sandy soil in a dry eucalypt forest was turned over by bandicoots and bettongs in their searches for fungi and invertebrates. However, populations of the main defoliators cannot be allowed to reach their potential sizes in the small reserves in the run country landscape, because of their easily realised preference for lusher pastures. In the runs themselves, sheep are preferred to native herbivores, for obvious economic reasons (see Chapter 2). Sheep are a cost-effective option for maintaining the bare ground required by many threatened species of the grasslands and grassy woodlands of subhumid Tasmania. For this reason, Tasmanian ecologists successfully argued for sheep grazing at the past low rates to be continued in the Tom Gibson Nature Reserve, and for the setting up of experimental plots to assess the effects of this grazing and the effects of its interaction with burning (Fensham 1991; Kirkpatrick, Gilfedder *et al.* 2005).

Dorrough, Yen *et al.* (2004), in questioning the similar 'status quo management' of the Terrick Terrick grasslands in Victoria (Diez and Foreman 1997; Foreman 1999), missed the point that such management is not advanced as a permanent panacea but only as the temporary best bet. It is the best bet because the species that have survived on the site have survived the same sort of management for many years, and it is usually not known how they would respond to other forms of management. The hypothesis that some species are experiencing a protracted death, and therefore need remedial action, can be tested. Exclosures that keep out sheep, but not natives, can give a good indication of the consequences of keeping sheep in the system at the status quo level and the rewards for taking them out.

A dramatic change in management has occurred with the declaration of the Mokota Reserve near Burra in South Australia. All stock were taken off this temperate grassland reserve. This change has proven to be very effective in producing powerful displays of the beautiful blue flowers of salvation jane (aka Paterson's curse), an exotic species eaten by sheep past the 'brussell sprouts' stage of selection. Stock removal has not proven effective in returning the grassland to a more native condition. In this somewhat Gallic case, sheep were assumed guilty of degrading natural values and their innocence tested by erecting enclosures stocked with sheep. In the Victorian and Tasmanian cases the principles of British law were followed, and guilt was tested by erecting exclosures.

The conclusions drawn from the results of observations based on exclosures or enclosures need to be cautious and based on appropriate criteria. In relation to the Terrick Terrick grassland exclosures (see Figure 5.5), Dorrough, Yen *et al.* (2004) argue that the facts that the seven-year record of change within the exclosures showed that the native annuals responded largely to climatic conditions, the diversity of native perennials increased and bare ground decreased: 'suggest that reductions in grazing intensity and frequency could improve landscape function and with little or no negative biodiversity effect' (p. 286). In relation to the biodiversity effect, they could have looked over the fence of the national park into the adjacent road reserve, where the successional consequences of little or no grazing are evident. These consequences are a dominance of large native perennials, many of which do not occur in the grazed areas, and an absence of small native annuals (see Figure 5.5). The combination of little or no grazing in the road reserve, and heavy grazing in the national park, results in the survival of more native grassland species than would be the case with either treatment on its own. It is nice to have landscapes functioning well, but not at the expense of the local, regional or global survival of native species.

Tongway and Hindley (2000) developed indicators that can be used to deduce soil stability, infiltration rates and nutrient cycling rates. These include soil cover, which relates to stability, basal area of perennial grasses, which relates to infiltration rates and nutrient cycling, litter cover, which relates to stability and infiltration rates, cryptogam cover, which relates to stability, erosion type and severity, which relates to stability, micro-topography, which relates to infiltration and nutrient cycling, and soil

Figure 5.5. Native grassland in stock grazed paddock at Terrick Terrick National Park, showing exclosure (above) and native grassland on adjacent ungrazed roadside (below) (Jamie Kirkpatrick).

texture, which relates to infiltration. Landscape functioning is assumed to be good if these and other indicators suggest a high degree of soil stability, high infiltration rates and a high rate of nutrient cycling. This valuation may well be appropriate when considering the consequences for physical sustainability of different treatments of agricultural or pastoral land. However, it does not work for nature conservation. There are large numbers of species, which, if they could verbalise their needs, would reverse one, two or all of the valuations. For example, there are many species of plant that require the soil disturbances that facilitate erosion, high infiltration rates do not favour species that require ephemeral ponds, and high levels of nutrient availability can kill species adapted to low nutrient levels. Using these criteria, many natural landscapes functioned very poorly well before humans evolved.

The admirable search for general principles of conservation management can be detrimental for some species, often the more threatened, because the conservation extension community is always in search of quick rules of thumb, and there are always exceptions to any rule. For example, after engaging in an appropriately named 'modified Delphi process', in which experts sought consensus on the optimum management of native grassy vegetation in south-eastern Queensland, McIntyre *et al.* (2000, p. 95) developed the following recommendation: 'Graze conservatively to maintain dominance of large and medium tussock grasses over 60–70% of the native pastures.' This was framed as a hypothesis, at the paddock scale, not the landscape scale, with the rationale that a level of grazing that allowed the natural spacing of tussocks to persist over 60–70 per cent of the paddock area would have enough closely grazed environments without tussocks to cater for those species that require such conditions, while ensuring habitat continuity for those that required taller vegetation. McIntyre and Tongway (2005) found that large tussock basal area was strongly related to all of stability, infiltration and nutrient cycling, as judged by the Tongway and Hindley (2000) indicators, which included perennial grass basal area, thus creating some minor degree of circularity in some of their analyses. However, the biodiversity implications of their threshold have not been tested.

As can be judged by the photographs in the present volume, most Tasmanian run country abjectly fails on the 60–70 per cent large tussock cover threshold, even though a 100 per cent cover of large tussocks can be produced in a year or two simply by excluding sheep (see Figure 5.6). Those native paddocks that fall most below this threshold tend to have high concentrations of those rare or threatened plant species that are tussock-averse. It is easy to maintain close to a 100 per cent vegetation cover on the Tasmanian runs without indulging in retaining a single tussock. The lack of palatability of large tussocks is the major reason why most Tasmanian graziers manage for lawns, with just the occasional group of individuals of sagg or tussock grass for shelter (refer to Chapter 2).

Experimentally informed adaptation from status quo management seems likely to be a better option than management from unsupported hypotheses or management from abstractions made faux concrete with complex formulae. What has been the experience with the status quo/experimental approach in the Tom Gibson Nature Reserve? The exclosures erected in the Tom Gibson Nature Reserve were placed in white gum grassy woodland (see Figure 5.6), black gum grassy woodland, and black peppermint heathy open-forest (see Figure 5.7), all in the same large paddock. The fences keep out sheep, but not the other herbivores. The white gum woodland had the most fertile soils and, by far, the greatest amount of grazing by sheep, rabbits and wallabies (Kirkpatrick, Gilfedder *et al.* 2005). The black peppermint heathy open-forest was at the other extreme. Half of each experimental site was burned, the more fertile two with great difficulty, because of a lack of fine fuel. The plots were set up and

Figure 5.6. A three-year-old exclosure, which excludes sheep, but not other herbivores, in grassy woodland in the Tom Gibson Nature Reserve, Tasmania (Kerry Bridle).

measured in 1996, then re-measured in 1999. After only three years it was apparent that there were not enough other herbivores in the reserve to keep the kangaroo grass sufficiently suppressed to allow the survival of the conservation-significant annuals in the white gum grassy woodland, burned or not (see Figure 5.6). At the less fertile and less grazed sites, sheep definitely had an effect on vegetation cover (Kirkpatrick, Gilfedder *et al.* 2005), but no species showed any indication of prospective local extinction from any of the four treatment combinations.

One other major management implication emerged from observations of these exclosures. In the white gum grassy woodland exotic herbs prospered most where there was burning but no sheep grazing (Kirkpatrick, Gilfedder *et al.* 2005). On the ungrazed Domain in Hobart, Kirkpatrick (1986a, 2004) observed an increase in exotic herb frequency with higher fire frequency, a phenomenon also observed by Lunt (1990) and Lunt and Morgan (1999) in Victoria. In the Spiky Bridge Coastal Reserve, which was once part of a native paddock that had some introduced fescue in the sward, the sheep-grazed lawn outside the reserve fence is dominated by native grasses and herbs, while the reserve vegetation is dominated by tall fescue. In Maria Island National Park, which has no sheep, but many native herbivores, exclosures in old paddocks quickly became dominated by exotic herbs and grasses. Yet, the grazed sward is dominated by natives. All this evidence suggests that sheep and wild herbivores prefer to consume introduced grasses and herbs where these are mixed with native grasses and herbs. It also supports the conclusions of Lunt (1990) and Lunt and Morgan (1999) that fire is not an adequate substitute for grazing in native grassland.

The practical implication of the results of the exclosure study for the management of the Tom Gibson Nature Reserve is that, unless native herbivore populations can be

Figure 5.7. A 10-year-old sheep exclosure in black peppermint (*Eucalyptus amygdalina*) heathy open-forest on the Tom Gibson Nature Reserve (Jamie Kirkpatrick).

allowed to get much larger, sheep grazing is necessary to maintain native biodiversity by reducing the biomass of the dominant native grasses, and may also be necessary to control exotic herbs and grasses. However, this conclusion pertains only to the small part of the reserve that is white gum grassy woodland. Sheep are totally unnecessary as conservation tools in the less fertile areas, although the numbers that are adequate to treat the white gum grassy woodland do not appear to present a danger to native plant biodiversity in the rest of the reserve. However, with heavier sheep grazing pressure than is experienced in the reserve, depletion of native species and invasion of exotic species could occur (Fensham and Kirkpatrick 1989; Fensham *et al.* 1999; Dorrough *et al.* 2004).

Fensham and Kirkpatrick (1989), who compared grassy vegetation on roadsides and in adjacent paddocks in Tasmania, observed that grazing is only beneficial for native plant biodiversity on the most productive sites. The similar studies of Fensham *et al.* (1999) in south-eastern Queensland and Dorrough *et al.* (2004) on the Monaro did not cover the same range of site conditions. This may account for their observations that native species richness was highest on the less disturbed of their sites, although other explanations cannot be excluded. The dynamics of grazing-environment interactions are obviously complex, as are the dynamics of fire-grazing interactions. Both of these are discussed in the next section.

Fire, air, water, earth and teeth

Kirkpatrick (2003) suggested that, in pre-European times, the major influence on vegetation patterns in the more fertile parts of Tasmania may have been herbivory, rather than fire, which was the dominant influence in less fertile parts of the state. Jackson (1968) developed a probabilistic model that successfully explained the relative distributions of rainforest, eucalypt forest, scrub and moorland on the poor soils of western Tasmania, a region with very high rainfall (Henderson and Wilkins 1975; Brown and Podger 1982). The essence of the model lay in a feedback between the flammability of the vegetation and the probability of ignition, with flammability decreasing with time elapsed since the last fire and the rapidity of the relay successional process, this rapidity being controlled by the potential productivity of the environment. Although Jackson (1968) mentioned that the outcomes of the model could be affected by concentrated grazing by vertebrates in small burned areas on more fertile soils, the model effectively ignored vertebrate grazing pressure.

In western Tasmania and other nutrient-poor, but moisture-rich, parts of the world, much of net primary production becomes potentially flammable litter, mor and wood, or unpalatable or poisonous foliage. Fuel accumulates and takes decades to equilibrate with breakdown. In fact, in the most nutrient-poor and moist parts of western Tasmania, accumulation can continue indefinitely, in the form of red fibrous mor peats under vegetation on well-drained sites, and in the form of muck peats on poorly drained sites (Bridle and Kirkpatrick 1997).

In subhumid climates on fertile soils, such as in much of eastern Tasmania, herbs, graminoids and grasses – life forms that native and exotic vertebrate herbivores find highly palatable – form major components of the vegetation. In parts of the natural landscape that are apparently capable of supporting shrub or tree species, intense grazing appears to have prevented their access, creating swards reminiscent of bowling greens, locally called lawns. These situations include dune swales (Kirkpatrick and Harris 1995); wetland margins (see Figure 5.8; Kirkpatrick 1975; Kirkpatrick and Harwood 1983b); and, outwash lobes beneath snow patches (Gibson and Kirkpatrick 1985).

As scat concentrations decline away from lawns, the density and cover of tussocks, shrubs and trees increases (Figure 5.9), these apparently requiring some abatement of grazing pressure to grow into an unpalatable state. Once established, they can reduce the palatable component of the vegetation, and therefore herbivore density, by lateral spread and root competition. Once the crowns of the canopies of tussock and shrub get close enough for fire to propagate between them, there is a potential for restoration of lawn, if grazing pressure is sufficient during the post-fire regeneration phase. Woody

Figure 5.8. Marsupial scats form a beach on a lawn formed around the edge of a wetland (Jamie Kirkpatrick).

plants that regenerate from epicormic shoots have the best chance of escape into unpalatability in the post-fire period, those that regenerate from rootstocks the next best. The pressure from grazing animals will depend on the scale of the fire, a small fire being likely to attract more grazing animals per unit area than a large fire. If there are large areas of palatable vegetation too short and green to burn, or unburnt by chance,

Figure 5.9. Unpalatable tussocks and shrubs on the edge of a marsupial lawn (Jamie Kirkpatrick).

larger marsupial populations will be carried over the gap between burn and regeneration than otherwise.

Because vertebrate herbivore populations tend to grow to maximum carrying capacity, irrespective of non-human predation pressure, as in the Kruger National Park (Eckhardt *et al.* 2000), the presence of adequate fuel in the ground layer to carry fire depends on one of the following circumstances: insufficient grazing pressure to prevent potentially inedible species attaining that state after fire because of a low carrying capacity between fires compared to the carrying capacity in the regeneration phase; increases in forage biomass related to climatic conditions that outstrip the potential increase in herbivore biomass through reproduction or migration; sudden density-induced mortality from epidemics among the herbivore population; high levels of sudden herbivore mortality related to extreme environmental events. In the run country of Tasmania, extensive fires often occur in the summer after a wet winter/spring that has been preceded by a series of dry years that have caused destocking and wild animal dieback.

On nutrient-poor soils, the difference between the nutrients that are available in palatable regrowth after fire and the nutrients that are available in palatable vegetation before fire is likely to be much larger than on nutrient-rich soils. Most of the nutrients in ecosystems on poor soils are held in unpalatable vegetation and only become available for palatable growth after fire (Bowman *et al.* 1986), whereas most of the nutrients on better soils are held in the soil and in palatable vegetation, whether in the regrowth or old growth state. This means that the poorer the soils are in nutrients the less likely it is that there will be sufficient grazing pressure to prevent palatable regrowth becoming unpalatable regrowth, a suggestion that fits patterns of woody plant persistence in the annually burned savannahs of the Kruger National Park (Mills and Fey 2005). An increase in fire frequency on poor soils is likely to progressively deplete the total ecosystem nutrient stock (Harwood and Jackson 1975; Bowman *et al.* 1986), while reducing the difference between pre-fire forage and post-fire forage. This will cause slowing growth rates, due to declining nutrient stocks, and increasing relative grazing pressure, thereby increasing the period between fires.

In environments with seasons in which forage growth is impossible, whether due to moisture deficits, cold or snow cover, there will be a period in which fine fuels can accumulate, because the vertebrate populations will adjust to the most forage-poor season. However, migratory animals can negate this fuel accumulation. In the winter in dry northern Australian savannahs, where migration is not a possibility, the density of native grazing animals is very low compared to areas of similar forage productivity in lowland eastern Tasmania, with its evenly distributed rainfall and high local relative relief. In the alpine zone of mainland Australia, the lack of accessibility to any forage over winter has resulted in an absence of large native vertebrate herbivores over the rest of the year. This is not the case in the Tasmanian alpine zone, where snow cover is not usually persistent over winter (Kirkpatrick 1986b; Bridle and Kirkpatrick 1998). However, years of persistent and heavy snow do result in substantial mortality of vertebrates (Jackson 1973).

It seems that fire frequency and intensity can be a partial function of grazing pressure.

A negative relationship between grazing pressure and fine fuel levels is highly likely to influence fire regimes. For example, Briggs *et al.* (2002) found that bison grazing increased the rate of invasion of woody plants in tall grass prairie, irrespective of fire frequency. They suggested that this phenomenon could be due to the patchy removal of fine fuel by the grazing animal, which was known to reduce fire intensity and extent (Knapp *et al.* 1999). In the *Pinus nigra* forests of the Pyrenees, a reduction in grazing has resulted in an increase in fuel and fire risk, largely through the agency of expansion of populations of *Genista* (Valderrobana and Torrano 2000). A role of domestic grazing animals in reducing fire frequency, and thereby promoting woody plant invasion, has long been posited (Madany and West 1983; Knapp and Soule 1998). Alternatively, herbivores might prevent woody plant invasion after fire by eating seedlings and resprouts. The nature of the relationship between grazing pressure and fire frequency will therefore depend on the palatability of woody species as much as the intensity of grazing.

Heavy grazing prevents potential fuel becoming inaccessible to grazers. Heavy grazing can maintain grass swards and tree and shrub resprouts in a constantly rejuvenating condition (Howe and Westley 1988) and can kill shrub and tree seedling regeneration (e.g. Cooke 1987). Young leaves tend to be less lignified and silicified, and more nutritious than older leaves (Howe and Westley 1988). Although grazing can result in selection for higher silica content, as well as dwarfing and asexual reproduction in grasses (McNaughton and Tarrants 1983; McNaughton 1984), grazed pastures can be more productive than ungrazed pastures (McNaughton 1976).

Some generalisations/hypotheses (shown in italics below) relevant to understanding the interactions of fire and grazing and their effect on vegetation dynamics in the runs can be deduced from the above discussion and the literature.

Fire increases the palatability of vegetation.	Young leaves tend to be less lignified and silicified, and more nutritious than older leaves (Howe and Westley 1988). Fire is generally not selective in its impact, and provides an environment suited to the regeneration of both palatable and unpalatable species (Bond 2005; Bond and Keeley 2005).
The more even the forage production through time the closer the animal populations will come to utilising it on an annual/decadal time scale, and the less will be the fuel available for a fire.	In environments with predictable seasonal extremes in forage production, non-migratory grazing vertebrates will have their adult numbers in the growing season largely controlled by resource availability in the poorest season. Most growth in the favourable season will form fuel, rather than being eaten. Inter-annual variation in conditions that influence productivity will have a similar, albeit damped, influence, the damping relating to a greater capacity for population recovery in longer time periods.
The ratio between the forage available in unburned vegetation and that available in the post-fire regeneration phase will influence the impact of vertebrate grazers on regenerating vegetation.	Soils of low nutrient availability tend to have lower densities of grazing mammals than those of high nutrient availability, because of low forage availability. When nutrients are liberated by fire, and become available to grazing animals in regrowth, such regrowth is likely to attain unpalatability before animal populations can build up sufficiently to keep it in a palatable state. In the more fertile areas supporting high densities of vertebrate herbivores, forage available in the post-fire regeneration phase is likely to be not so dramatically different from the long unburned state, resulting in a greater post-fire impact of herbivores on regeneration.
The spatial extent and patterning of burns will influence the impact of grazing on fire regimes.	Small areas of burned vegetation will result in a concentration of herbivores, while large areas will have lesser concentrations (Jackson 1968).

The above hypotheses imply that grazing and fire together will have greater effects on vegetation than the sum of their single effects, as was supported by the results from the Tom Gibson Nature Reserve exclosures, where interaction effects occurred on all but the least fertile site (Kirkpatrick, Gilfedder *et al.* 2005). Synergistic effects could be expected to be greatest in ecosystems rich in palatable grasses and herbs, as was the case with the Tom Gibson data. The role of grazing in reducing fire frequency may be critical in the vegetation dynamics of the runs, as is strongly indicated by data presented in the previous chapter. The small patch burning of unpalatable vegetation that is used by many Tasmanian graziers (see Chapter 2) concentrates grazing and thereby causes increased palatability, as could be expected from some of the above hypotheses.

Effects of grazing regimes on plants

There have been several investigations of the influences of different grazing regimes on the relative abundance of species and life forms in the Tasmanian runs (Friend, Dolan *et al.* 1999; Garden *et al.* 2000; Leonard and Kirkpatrick 2004; Kirkpatrick, Bridle *et al.* 2005; Kirkpatrick, Gilfedder *et al.* 2005). Spring spelling seems the most productive and conservation-friendly of the many alternatives that have been tested. However, experiment and observation have been located only in unfertilised areas with low rates of stocking, so inferences from them may not pertain to other situations. There is little indication in any of the results that changes in the timing and duration of grazing are likely to result in the loss of native plant species, although there is enough variation in the relative abundances of species under different regimes to suggest that spatial variability in management regimes might be a good thing in maintaining landscape-scale plant biodiversity.

Despite the lack of any evidence of potential local extinction of native plant species as a result of a change in grazing systems, there are theoretical grounds for suspecting that the more extreme of the rotational systems, such as cell grazing, might select against annuals and upright shrubs, by creating an alternation of a brief period of total defoliation of almost all plants with a longer period of establishment of complete cover in the cell as a whole. Annuals may disappear because periods of defoliation are controlled by the amount of forage, and, therefore, do not necessarily directly precede the incidence of conditions suited to their germination. Germination may occur when a dense cover prevents establishment, or when mobs of sheep are in the cell. In this latter case the annual could be eaten out before setting seed for the next round. Shrubs below sheep head height, that might have survived set stocking because of low palatability, seem likely to be defoliated too frequently under cell grazing to survive in the long term. These are both slow and erratic processes, especially given the confusing effects of variation in climatic conditions between years, and demonstrate the need for granting bodies to fund projects for longer than the usual three to five years.

Sheep and invertebrates

Sheep eat invertebrates only as by-catch. However, they provide part of the resource base for some invertebrates, and influence vegetation structure and composition, thereby indirectly affecting invertebrate assemblages. The use of sheep by green blowflies has been documented earlier. The maggots of the green blowfly are occasional victims of the maggots of two species of native blowfly, which sometimes get blown over Bass Strait from the mainland (Peter McQuillan, personal communication, May 2006). In Spain, sheep dung supports a diverse collection of coprophilic beetle species, 13 taxa being collected by Lobo *et al.* (2006) from one site. In contrast, in Tasmania, sheep scats tend to sit until broken down by the elements.

Mysterud *et al.* (2005) undertook a well-replicated landscape-scale experiment to determine the impacts of different intensities of sheep grazing on invertebrates in Norwegian natural vegetation. At the end of the first year there was no effect of sheep grazing intensity on invertebrate composition or richness. Rather, variation in the invertebrate communities was largely related to plant species composition and vegetation structure. One year was not enough time for variations in stocking to cause changes to the environmental attributes that most affect invertebrates.

Many attributes of the environment, and invertebrate fauna, of the runs can be changed by longer-term increases in sheep grazing pressure. Soils are compacted by stock (Kirkpatrick *et al.* 2000). Such compaction is known to impact on soil fauna (Hutchinson and King 1970; Cluzeau *et al.* 1992). Reductions in litter cover can result from heavier stocking by sheep (Kirkpatrick, Gilfedder *et al.* 2005). This is likely to have consequences for detritivores (King *et al.* 1976; King and Hutchinson 1983; Abensberg-Traun *et al.* 1996). Increased sheep grazing pressure tends to make grassy vegetation more lawn-like, by eliminating, or reducing the stature of, tussock-forming species. This transition may favour invertebrate species that cannot eat silicified older foliage, such as grasshoppers (Tschamtke and Greiler 1995), but eliminates habitat for many others, such as orb-weaving spiders (Gibson *et al.* 1992) and species that have evolved to feed on tussock seeds or foliage (McQuillan 1999). The larvae of the rare ptunarra brown butterfly depend on *Poa labillardierei* tussocks for their sustenance in the run country of Tasmania (McQuillan 1999).

The effects of sheep grazing on invertebrate assemblages have been studied on two extreme sites in runs in Tasmania, the high altitude grassy shrublands of the Central Plateau and the Tom Gibson Nature Reserve in the lowlands. The exclusion of sheep for more than 20 years from an exclosure on the Liaweenee Moor resulted in the development of large *Poa* tussocks, absent from the surrounding sheep-grazed country (refer to Figure 1.2), and a marked increase in cover and diversity of native forbs (Bridle and Kirkpatrick 1999). Francis (1997) found that, in three months sheep ate 50 per cent of the biomass of tussocks outside the exclosures, accounting for their small stature. He also found that tussocks of the size induced by sheep grazing were bereft of the invertebrate fauna found on the larger tussocks within the exclosure. The number of families of invertebrates collected from individual tussocks tended to increase with their size. The largest tussock contained individuals from 12 families. In sampling invertebrates from the vegetation as a whole inside and outside the exclosures, he found that the sheep-grazed vegetation had a higher richness of families that utilised fungi or pollen, and that the vegetation inside the exclosures had a higher richness of families that fed on grasses and forbs.

Fitzgerald (2004) collected and identified invertebrates from the white gum grassy woodland and black peppermint heathy open-forest experimental sites in the Tom Gibson Nature Reserve. Invertebrate taxon composition proved to be strongly influenced by all of vegetation type, the presence or absence of grazing, the presence or absence of burning and the interaction between vegetation type and burning. Taxon richness responded to burning and the interaction of burning with grazing, with richness being higher in burned and grazed plots than in other combinations of treatment. Of 142 taxa, 124 were found at the black peppermint site and 113 in the white gum site, with 77 per cent of taxa being in common. However, only one species which had a total abundance of greater than 20, and a frequency in the pitfall traps in one vegetation type of over 50 per cent, was confined to one vegetation type. There were few similarly abundant taxa that were confined to either of the burning or grazing treatments. Therefore, although burning, grazing and vegetation type strongly influenced invertebrate assemblages, this was more to do with relative abundances, rather than local extinction, as with the response of plant species to different management regimes. However, the same result might not have been gained if the combinations of treatments were not in such close proximity.

Sheep and native vertebrates

Sheep form part of the diet for many carnivorous native animals. The useful role of the Tasmanian devil in reducing sheep carcases to wool balls has already been noted. The nationally threatened spotted-tailed quoll (*Dasyurus maculatus*) has also been observed in bush runs in the Midlands (I. Skira, personal communication, 2003), where it almost certainly indulges in the occasional dead sheep. The eastern quoll (*Dasyurus viverrinus*), a species believed to be extinct on mainland Australia, is common in the run and front country, where it consumes corbie grubs and cockchafers, among slightly larger prey and carrion. The Tasmanian wedge-tailed eagle is another conservation-significant native carnivore that is not averse to consuming the occasional live or dead lamb.

Guides for the maintenance of habitat for native vertebrate animals tend to stress the importance of coarse woody debris, old-growth live trees, dead trees, large tussocks, deep litter and dense, prickly shrubberies (Lindenmayer *et al.* 2003; Mokany *et al.* 2006). Most of these characteristics would be regarded by graziers as symptoms of badly managed bush runs. As with vascular plants and invertebrates, vertebrates vary in their habitat preferences. Some like dense, prickly shrubberies, some like dense, prickly shrubberies in moderation, and others cannot use them. Bird species provide good examples of this range of responses to dense, prickly shrubs (MacDonald and Kirkpatrick 2003; see also Chapter 6). There are many native vertebrate species that occur in the runs that require dense ground shelter. The southern brown bandicoot (*Isoodon obesulus*) is one. The tussock skink (*Pseudemoia pagenstecheri*) is another. Hutchinson *et al.* (2001, p. 35) remarked that this rare reptile of low rainfall grassland and grassy woodland: 'shelters inside the bases of tussocks and basks inconspicuously in the spaces between them.'

Sheep grazing regimes that produce lawns therefore seem likely to disadvantage a substantial subset of native vertebrate species that occur on the runs. The persistence of species such as the southern brown bandicoot and tussock skink may rely on local heterogeneity in management regimes and environment. In recent years this heterogeneity has been increased by the widespread fencing off of parts of bush runs and streams and an increasing adoption of rotational systems. On the other hand, rotational systems tend to reduce local heterogeneity.

Conclusions

There is no right way to manage sheep grazing to promote the interests of 'nature' on the runs. Any management regime will benefit some native species and disadvantage others. All of the regimes currently used on the runs can provide habitat for at least some of the many rare or threatened native species that coexist with sheep, or are benefited by them. One key to maintaining native biodiversity in the runs as a whole is the maintenance of spatial heterogeneity in management at the landscape scale. The management regimes that allow the grassland paperdaisy to persist are very different to those required by the tussock skink. In contrast, it seems likely that temporal heterogeneity in management will lead to the loss or decline of species that were adapted to the old management regime, but not the new.

Conclusion

Chapter 6

Run country on the run

J.B. Kirkpatrick, L. Gilfedder, L. Mendel and E. Jenkin

Summary

Between 1972 and 1999 the clearance of Tasmanian runs for crops, improved pasture and dams declined dramatically, although, since then, there has been clearance of some of the least-protected vegetation types. During a six-year period in the 1990s some vegetation remnants in wool-growing country were cleared or partially cleared and a substantial number of properties changed ownership. Change in ownership was often associated with changes in management of the remnants. Some changes were positive; more were negative. The vegetation characteristics of the edges and centres of the uncleared remnants did not change very much over the six years. Edge effects in remnants are shown to be very limited in penetration and largely related to ingress of nutrients from surrounding developed country. Mosses and liverworts, vascular plants and birds respond very differently to the characteristics and management of remnants. Small remnants in poor condition may, in some cases, be more important for the maintenance of native biodiversity than large remnants in good condition. The management of remnants for nature conservation should be primarily directed towards the maintenance of populations of threatened species, not improvement in condition.

Introduction

Some of the ecological effects of management of runs for wool production have been described in the previous four chapters. While the effects of variation in management of the runs are important for nature conservation, they pale into insignificance in comparison to the effects on native organisms of the conversion of runs to improved pasture, cropland, tree plantations and water storages. Such clearing results in the loss of most native species, and their replacement by exotics or water. While clearing is a

Figure 6.1. A landscape in run country, with some small areas of partial clearance (Jamie Kirkpatrick).

Figure 6.2. The rural landscape around Campbell Town, where few remnants of native vegetation survive (Jamie Kirkpatrick).

Figure 6.3. A variegated landscape in south-eastern Tasmania, with fire (Jamie Kirkpatrick).

quick process, restoration of native vegetation is slow and uncertain in its consequences, whether carried out by human beings or left to natural processes. Therefore, clearance is the main source of potential conflict between nature conservation and wool production, justifying an effort to understand its causes, incidence and nature conservation consequences.

The pattern of cleared and native country in the wool-growing areas of Tasmania varies from landscapes that have only the occasional patch of semi-cleared land (see Figure 6.1) to landscapes in which there are only occasional small remnants of native vegetation (as shown in Figure 6.2). In between these two extremes there are variegated landscapes (McIntyre and Barrett 1992), in which cleared land, partially cleared land, cleared land being reinvaded by natives to varying degrees, native land which has lost its trees and native land that is almost in pre-European condition, occur in complicated mosaics, partly related to their productive capability (see Figure 6.3). Because the gently sloping, more fertile land has been the most cleared, species and plant communities that were confined to this land have become rare or threatened, making native vegetation remnants in this type of country very valuable for nature conservation. For this reason it is important to understand the ecology of these remnants, how they respond to various management regimes, and their viability in the long term. Thus, the subject matter of this chapter is the nature and consequences of the clearance of run country (see map of vegetation clearance, Figure 6.4). The potential for cleared country to revert to native vegetation is also briefly covered.

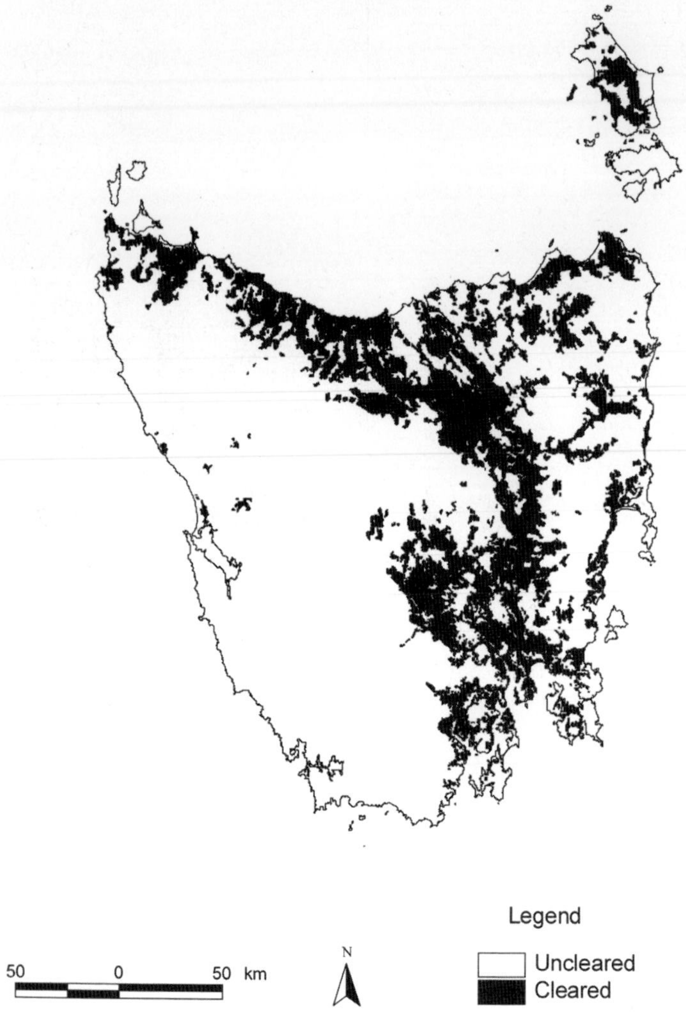

Figure 6.4. Clearance of native vegetation in Tasmania 1802–1972.

Clearing

The second half of the twentieth century was an era of massive conversion of runs to improved country (refer to Chapter 1). Since 1972, when satellite images became available for Tasmania, it has been possible to monitor the patterns of land clearance. Between 1972 and 1980 large areas of the run country were cleared, particularly on properties in the medium elevation country (Kirkpatrick and Dickinson 1982). This pattern continued, less rapidly, in the period 1980–1988 (Kirkpatrick 1991). To follow up on this previous work, the extent of land clearance in Tasmania from 1988 to 1999 (see map, Figure 6.5) was also interpreted from satellite images.[1]

The overall pattern of land clearance in Tasmania between 1972 and 1999 is a mixture of attrition of bush for pasture at the edge of the improved land, and insertion

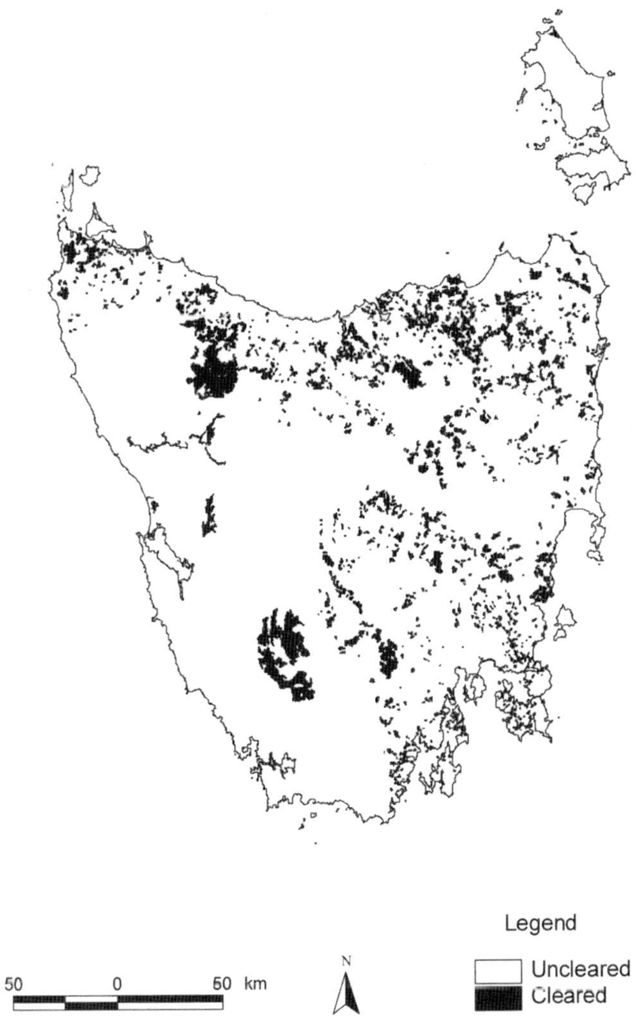

Figure 6.5. The distribution of land clearance in Tasmania 1972–1999.

of dams and plantations within large areas of bush (refer to Figures 6.4 and 6.5). The two major vegetation types that have been cleared on wool-growing properties are dry eucalypt forest and grassland; their rates of loss were high both before and after 1972 (see Table 6.1). In the period 1980–1999 the rates of loss steadily decreased for sclerophyll forest, grassy woodland, coastal grassy forest, inland grassy forest and grassland, the vegetation types most common on the runs, and steadily increased for rainforest, *Eucalyptus delegatensis* tall forest, *E. obliqua* tall forest, *E. nitida* scrub, wet scrub, *Allocasuarina* forest and heath (refer to Table 6.2), largely reflecting clearance for agriculture in the north-west and the Furneaux Group of islands and the increasing establishment of timber plantations in areas of high rainfall.

The mean annual native vegetation loss declined from 10 429 ha in the period 1988–1994 to 6992 ha in the period 1994–1999. Clearance for agriculture declined

Table 6.1. Mean percentage loss per decade of major vegetation types in Tasmania for 1802–1972 and 1972–1999, showing estimated area in 1802

Vegetation type	Area in 1802 (ha)	1802–1972	1972–1999
Wetland	24 500	1.75	3.52
Dry eucalypt forest	2 853 300	2.43	2.04
Moorland and scrub	1 150 000	0.05	1.41
Wet eucalypt forest	1 272 800	1.27	1.27
Grassland	85 000	2.12	1.01
Rainforest	647 600	0.69	0.59
Heath	425 800	2.49	0.52
Swamp forest	39 000	3.49	0.31
Saltmarsh	4000	0.90	0.01
Alpine	115 000	0.00	0.00

from an annual rate of 5045 ha in 1988–1994 to 2879 ha in 1994–1999. Clearance for plantations slowed down slightly from 4796 ha to 4113 ha per annum between the same periods. There were no new impoundments in the latter period, whereas 587 ha of native vegetation per annum were inundated between 1988 and 1994.

The tendency for vegetation loss to shift from lower to higher rainfall areas that was evident in comparing 1980–1988 to 1988–1994 (see Table 6.2) intensified between 1994 and 1999. Between 1994 and 1999 there was little clearance in the dry Midlands and far north-east of the state, major regions of clearing between 1972 and 1988. After 1994, large amounts of native vegetation were replaced by farmland and tree plantation in the north-west of the state, and there was considerable clearance of bush for pasture on Flinders Island.

There are several possible explanations for this change in the pattern of native vegetation loss. Plantations require moist sites to produce growth rates that justify investment. They have been subsidised to replace production from reserved native forests, and have been taxed more favourably than clearance for agriculture. Retail managed investment products have been popular, leading to tree plantation establishment in progressively drier areas in the first decade of the twenty-first century.

The returns from wool declined markedly during the 1990s, giving less incentive and capability to clear the drier country, which is mainly used for sheep grazing. The economic situation was less dire in country used mostly for dairying or beef production. The availability of bush suitable for clearing was not likely to have been a limiting factor; although many wool-growing properties had no such bush left, others had substantial amounts.

The impacts of land clearance on nature conservation values declined steadily between 1980 and 1999. The vegetation types most cleared in the late nineties were well represented in reserves. The forest types of greatest conservation significance were

Table 6.2. Clearance (ha) by cause and vegetation type, 1980–1999

Vegetation type	PA	PP	PS	Total	Area	% loss p.a.		
						1994–1999	**1988–1994**	**1980–1988**
**E. delegatensis* tall forest	157	4387	3799	8343	330 100	0.51	0.07	0.00
Heath	2815	0	0	2815	146 481	0.38	0.12	0.03
**E. obliqua* tall forest	2119	1109	5756	8984	549 600	0.33	0.15	0.03
*Swamp forest	138	0	0	138	9449	0.29	0.79	0.26
Sclerophyll forest	1696	94	191	1981	179 200	0.22	0.57	0.77
Grassy woodland	512	0	0	512	50 394	0.20	0.21	0.74
**E. obliqua* wet forest	1066	395	170	1631	183 100	0.18	0.92	0.08
*Rainforest	132	2948	728	3808	606 549	0.13	0.10	0.08
Coastal grassy forest	683	0	255	938	192 203	0.09	0.34	0.42
Allocasuarina forest	19	0	0	19	4228	0.09	0.04	0.00
*Wet scrub	298	12	129	439	116 395	0.08	0.07	0.04
*Buttongrass moorland	3775	0	0	3775	965 835	0.08	0.10	0.03
E. delegatensis forest	507	345	85	937	334 100	0.06	0.64	0.03
Inland grassy forest	364	0	106	469	174 352	0.05	0.54	0.65
E. sieberi forest	100	0	0	100	48 082	0.04	0.19	0.09
**E. nitida* scrub	13	56	0	69	59 933	0.02	0.01	0.00
Inland *E. tenuiramis* forest	0	0	0	0	59 536	0.00	0.80	0.66
Grassland	0	0	0	0	61 799	0.00	0.18	0.20
Montane grassy forest	0	0	0	0	37 124	0.00	0.09	0.00
Coastal *E. tenuiramis* forest	0	0	0	0	12 810	0.00	0.03	0.00
Total	**14 394**	**9346**	**11 219**	**34 960**				

Vegetation types follow Kirkpatrick and Dickinson (1984). Mean annual % loss by vegetation type on the basis of surviving area in 1988 (Kirkpatrick *et al.* 1995) for 1988–1994 and 1994–1999 and surviving area in 1980 (Kirkpatrick 1991) for 1980–1988. PA = private land, agricultural use; PP = private land, plantation use; PS = public land, plantation use. * = vegetation type found largely in areas with > 1000 mm per annum of precipitation.

largely protected from clearing through regulations under the *Forest Practices Act 1985* (Tasmania). Nevertheless, clearance continues in the twenty-first century in vegetation types that have been highly depleted in the past and are poorly represented in the conservation estate. The vegetation types of most concern in this respect are grasslands and grassy woodlands. The current moves towards control of clearance of non-forest vegetation will be discussed, in the context of the future of the runs, in Chapter 7.

Natives invade previously improved land

The process of clearing replaces native plants with exotic plants by direct destruction of the natives using various combinations of felling, firing, ploughing and poisoning through the use of herbicides, and by changing the environment to better suit the desired exotic plants than the natives, the two major mechanisms being drainage and fertilisation. Natural levels of extractable phosphorus in soils under native vegetation are extremely low compared to those of soils under improved pasture and cropland. Only 8 out of 180 surface soil samples in native vegetation remnants in the northern Midlands had values over 20 μg/g, with only one having a value over 50 μg/g.[2] Most

Figure 6.6. Ruins of a nineteenth century house on a run, covered by native plants (Jamie Kirkpatrick).

were less than 10 μg/g. These figures are much less than those typical of healthy improved pasture, which tend to be over 40 μg/g (Zacharek 1997). In consequence, most native species have a competitive disadvantage. The only exception in the analyses of Kirkpatrick, Gilfedder *et al.* (2005) was the largest species of tussock grass (*Poa labillardierei*), which grew better with higher nutrient levels. In contrast to most of the natives, the exotic thistles, ryegrass and clovers flourished under fertilisation (Kirkpatrick, Gilfedder *et al.* 2005).

The increase in nutrients available to plants that fertilisers produce is transient. Nutrients become bound up in the soil in unavailable form and are exported in farm produce and run-off. Eventually they fall to levels that better suit the natives than the more nutrient-demanding of the exotics. Very few of the grassland and grassy woodland native species have a long-term soil seed store (Gilfedder and Kirkpatrick 1993b; Friend, Cameron *et al.* 1997), so recolonisation of such pastures largely depends on dispersal from native vegetation remnants. Some plants, such as eucalypts, have very limited dispersal ability. Almost all eucalypt seed falls within twice tree height. Others, such as many daisies, can travel long distances on the wind. Many species, such as the wallaby and spear grasses, are dispersed by mammals. Others, such as many wetland plants, and those with succulent fruits, are dispersed over moderate distances by birds. This means the process of recolonisation by natives can be slow and unbalanced if the only native vegetation remnants are far away. Nevertheless, at a decade timescale, it does occur; in the run country landscape there are many examples of ruins of houses and gardens, and nineteenth century mound-ploughed fields, covered by an assemblage of native plants as rich as those in areas that have never been modified (Figure 6.6).

Changes in native vegetation remnants 1993–1999

Introduction

Land clearance has left in its wake many islands of native vegetation in a sea of cultivation. Most of these islands are small. A few are very large. In 1993, 100 remnants in subhumid Tasmania were surveyed. They were in the range 5–200 ha, received less than 700 mm mean annual rainfall and were below 500 m above sea level, making them all in the subhumid temperate climatic zone. Remnants that were visibly extremely degraded were not selected. Data were collected on remnant flora, vegetation, geometric characteristics, age and management (Gilfedder and Kirkpatrick 1993a, 1995b, 1998a; Kirkpatrick and Gilfedder 1995). Most of these remnants were resurveyed in 1999,[3] in order to determine their survival, changes in their condition, changes in management and changes in ownership in a period in which there had been a large amount of activity in the region directed towards remnant vegetation conservation but no regulation of land clearance or management. If incentives, education and extension, in the absence of regulation, were effective means of conserving significant vegetation remnants there should have been no loss of remnants or any deterioration in their quality. We also wished to determine the effects of changes in ownership on the survival and management of remnants, the role of remnant characteristics in influencing changes in management, and, the nature of the vegetation changes that occurred in the surviving remnants.

Changes in ownership and management

Of the 85 re-accessed private remnants, 19 had experienced a change in ownership in the period 1993–1999 (see Table 6.3 below). The nine remnants in public ownership in 1993 remained in this tenure in 1999. Between 1993 and 1999: four of the accessed remnants had been cleared for agriculture; seven had their area reduced by clearing for roads, houses, agriculture or parking areas; three had management that was less likely to promote conservation values than the management in 1993; 63 had highly similar management; and 17 had improved management (as shown in Table 6.3). Only two of the nine publicly owned remnants did not have constant management. Part of the Hobart Domain remnant was cleared for parking, although weed control improved in the remainder. The Gorge/Trevallyn State Recreation Area remnant in Launceston had improved weed management.

All the remnants that were cleared or had poorer management had the same owners through the time period, while five out of the six private remnants that suffered depletion in area had changed ownership in the time period (see Table 6.3). Twenty-six per cent of the remnants that had a change in private ownership between 1993 and 1999 had been cleared, partially cleared or more poorly managed, compared to 12 per cent of remnants with the same private owners. Remnants with better management differed little between those with changed private ownership (16 per cent) and constant private ownership (19 per cent).

The fact that over one-fifth of the 85 private remnants changed ownership during the six years covered by the study indicates that any remnant conservation strategy that assumes constancy of ownership is flawed. Changes in ownership could logically be expected to increase the chances of changes in management, either for the better or for worse. There is a weak indication of this in the data set with 68 per cent of private remnants with the same owners at both dates having constant management compared

Table 6.3. Changes in the management of remnants between 1993 and 1999 by type of ownership and ownership change

	Private, no ownership change	Private, ownership change	Public	Total
Cleared	4	0	0	4
Reduced in area	1	5	1	7
Poorer management	3	0	0	3
Same management	45	11	7	63
Better management	13	3	1	17
Total	66	19	9	94

to approximately 58 per cent with different ownership. Changes in ownership of private remnants could also be expected to lead to a greater chance of deterioration or destruction of remnants, with the replacement of a conservative older generation by either the younger generation or new owners who need to develop land in order to service debt, or put their mark on the property. Our data are consistent in tendency with this expectation, while not being statistically significant (Chi squared, $P > 0.05$). Change in ownership did not affect the chances of improvement in management.

The intensive education, extension and incentive activity enabled by Natural Heritage Trust and other funding over most of the time period covered by this study is likely to have contributed to the improved management in 18 per cent of the remnants. However, the rate of attrition (especially in the context of rapidity in change of land ownership) indicates that education, extension and incentives are insufficient in themselves to ensure the medium-term future of remnants, a conclusion also reached by Gilfedder and Kirkpatrick (1997) on the basis of a social survey of managers of native grassland.

Change in management relative to remnant characteristics

This section asks whether any of the characteristics of the remnants (see note 4 for a list) made them more susceptible to positive or negative management changes. There were few such relationships. The soil pH in the remnant centre was lower ($P < 0.05$) in remnants with improved management than in those with constant management and those with poorer management/area loss (refer to Table 6.4). The concentration of improved management on remnants with low soil pH may relate to a preference among landowners for active conservation of bush on poorer soils, where the opportunity costs of such management are lowest. The concentration is also consistent with the natural human tendency to improve management where the symptoms of degradation are most apparent. If nutrient drift occurs, the vegetation of the sandy soils, which have the lowest pH values, seems more susceptible to exotic invasion than the vegetation on soils with higher pH values. Partially supporting this idea, available P on the remnant edge, but not the middle, was greater in those remnants with improved

Table 6.4. Significant differences in mean remnant characteristics (measured in 1993) between those remnants with constant area and management between 1993 and 1999, those remnants that were more poorly managed or partially or totally cleared, and those remnants with improved management (one-way ANOVA)

	pH centre	P edge	Per./area	Ex./Nat.
Poorer management/ loss of area	5.69a	11.1ab	1.61a	1.07b
Constant management/area	5.62a	8.6b	0.92b	1.33b
Improved management	5.25b	18.8a	1.38a	4.69a
Probability	0.03	0.04	0.04	0.03

Values that are not significantly different ($P > 0.05$) between rows share a letter (two sample t-test). Per. = perimeter, Ex./Nat. = the ratio between exotic and native cover on the warm edge of the remnant.

management than those with constant management, and the ratio between exotic and native cover on the warm edge, but not the middle, was higher in remnants with improved management than in other remnants (see Table 6.4). Remnants with poorer management/loss and those with improved management had higher perimeter/area ratios than those with constant management (see Table 6.4). It may be that the less compact remnants tempted either remediation or truncation.

The major implication of the above results is that the education, extension and incentive activity in the 1990s resulted in improvements in management that were biased towards remnants on land of lower productive value. These are not necessarily the remnants that are most important for nature conservation, as it has been the species and vegetation types on the most productive land that have been most depleted by past clearing.

Changes in vegetation characteristics

The remnants that survived changed little in the vegetation characteristics that do not change markedly with short-term changes in climate. The ground covered by tree trunks (as indicated by a Bitterlich wedge count), shrub cover and native grass cover, exhibited no change between 1993 and 1999 (see Table 6.5). Variables with a high short-lived plant component, exotic grass cover and herb cover, exhibited decline (refer to Table 6.5). These declines, and the increase in lichen cover, are consistent with the dry conditions that prevailed during data collection in 1999 compared to 1993. The increase in litter cover and decline in bare ground cover may also have had an influence on the success of short-lived herbs and grasses. The increase in litter cover, and

Table 6.5. The significance and direction of changes in cover classes between 1993 and 1999 at the cool edge, centre and warm edge of remnants (paired t-test, $P < 0.05$)

	Cool	Centre	Warm
Basal area (Bitterlich wedge)	NS	NS	NS
Native grass	NS	NS	NS
Exotic grass	NS	NS	**
Native herb	***	***	**
Exotic herb	NS	NS	***
Native shrub	NS	NS	NS
Exotic shrub	NS	NS	NS
Lichen	*+	***+	***+
Moss	NS	NS	NS
Exotic cover/native cover	NS	NS	**
Rock	NS	NS	NS
Bare soil	**	***	NS
Litter	***+	***+	***+
Total unvegetated	NS	NS	***+

+ indicates an increase. NS = $P > 0.05$, * = $P < 0.05$, ** = $P < 0.01$, *** = $P < 0.001$

decrease in bare ground, could reflect a decline in the use of burning and grazing during the time period. This tendency may have negative implications for many of the rarer native herb species (Gilfedder and Kirkpatrick 1994ab, 1998b; Kirkpatrick and Gilfedder 1995, 1998).

Do size and age matter? Edge effects in remnants

Introduction

Accepted ecological wisdom and deductions from theory have not been demonstrated to be a reliable guide to action in remnant conservation in Australia when ecologists address reality. Kirkpatrick and Gilfedder (1995) and Woolley and Kirkpatrick (1999) found that vegetation condition was not a predictor of importance of remnants for rare or threatened plant species. Prober and Thiele (1995), Gilfedder and Kirkpatrick (1998a) and Woolley and Kirkpatrick (1999) found that age and geometric variables had little influence on native vascular plant species richness and cover in remnants compared to the type of management. The management regimes that best maintained condition varied markedly between the different vegetation types that were covered by these studies.

Edge effects are an important area of remnant conservation ecology in which prevailing ecological wisdom (Saunders *et al.* 1991) merits testing against the contingencies of the real world. The ratios between exotic species cover and native species cover and exotic species richness and native species richness are employed as indicators of vegetation condition in the edges and centres of the 100 subhumid Tasmanian remnants described in the previous section. The information is derived from three sites in each remnant: the centre; the warm edge, usually the north-western extreme; and, the cold edge, usually the south-eastern extreme. In this section of the chapter variation in vegetation condition and soil nutrient status between the centres and the edges of remnants is described in order to determine the prevalence and magnitude of edge effects. Age, and the environmental, management and geometric attributes of remnants, are then related to the contrasts between centres and edges. Details of the methods used to produce the results in this section can be found in note 4.

Vegetation differences between edges and centres

There were differences between the centre and the edges in the ratios of exotic/native cover and exotic/native richness, but not between the cold and the warm edges (see Table 6.6). The centre significantly differed from both the cold and warm edges in exotic grass cover, exotic herb cover, native shrub cover and litter cover (refer to Table 6.6), with the former two variables being higher in cover at the edges and the latter being higher in cover in the centre. There were no significant differences between the two edges. The ratios of exotic and native cover and exotic and native richness had higher values than the centre in remnants with a width of less than 600 m. In the wider remnants there was no difference between the edges and the centre.

Table 6.6. Means for the centre, cold edge and warm edge in exotic/native richness and cover, cover types and soil characteristics

Variable	Warm edge	Centre	Cold edge
Exotic/native richness	0.31b	0.20a	0.29b
Exotic/native cover	2.01b	0.45a	1.27b
Native grass cover (%)	32.2a	32.2a	29.5a
Exotic grass cover (%)	16.4b	7.3a	17.2b
Native herb cover (%)	3.6a	4.7a	4.0a
Exotic herb cover (%)	11.1b	4.6a	10.9b
Native shrub cover (%)	4.6a	8.6b	3.2a
Exotic shrub cover (%)	1.8a	0.3a	1.5a
Fern cover (%)	0.3a	0.5a	0.6a
Moss cover (%)	3.8a	5.5a	4.1a
Lichen cover (%)	0.3a	0.3a	0.1a
Bare ground (%)	7.2a	7.2a	9.6a
Litter cover (%)	17.0a	26.1b	18.5a
Rock cover (%)	1.6a	2.9b	1.3a
pH	5.6a	5.6a	NA
Conductivity	0.08a	0.07a	NA
Available P (ppm)	10.9b	6.9a	NA
K (ppm)	202a	205a	NA
N (%)	0.26a	0.26a	NA

Values with the same letter in rows are not significantly different ($P > 0.05$, paired t-test).

These results confirm the existence of edge effects in the Tasmanian subhumid remnants, except for the widest of the remnants. Our data do not readily allow calculation of the depth of edge effects, but do allow the conclusion that they do not generally penetrate to the centre of the narrowest remnants. The absence of edge effects in the largest remnants is likely to be a product of their greater susceptibility to disturbance in their centres than the smaller remnants, with many of the remnants with the greatest widths having roads and developments of various sorts in their middles. This indication of a lack of penetration of edge effects is strongly supported by the work of Loofs-Samorzewski (2003) in heathy ecosystems in Tasmania. She found that edge effects penetrated only a few metres and that the effects of nutrient drift from developed land were minimal beyond this narrow margin.

Soil differences between edges and centres

Of the soil characteristics, only P was different between the centre and the edge, being higher on the edge (see Table 6.6). The lack of differences in N, K and pH between the centre and the edge indicates that N/P/K fertilisers are not widely used on adjacent land. Nitrogen is most likely to be used for cropping and pastures for dairy cattle (A. Cameron, personal communication, 2006). Potassium is used on new ground or hay paddocks but is not repeatedly applied as it is retained in the nutrient cycle and not depleted by livestock grazing (A. Cameron, personal communication, 2006).

Effect of remnant age on differences between edges and centres

The age of a remnant is the number of years since its present boundaries were established by clearing. The theory of island biogeography (MacArthur and Wilson 1967) suggests that the number of species on any island is controlled by the balance between extinction and immigration of species. In remnants, the species that become extinct are more likely to be natives than exotics, while the species that immigrate are more likely to be exotics from the surrounding matrix than natives from other remnants. Thus, one would expect that the older a remnant the more it would be dominated by exotic species, and the less would be the difference between the centre and the edges. However, there were no effects of age of remnant on any of the cover or richness ratios. These results are consistent with a lack of any clear effect of age on the condition of the same remnants as a whole (Gilfedder and Kirkpatrick 1998a) and on lowland basalt remnants in northern Tasmania (Woolley and Kirkpatrick 1999). The only relationships between age and other variables included in this study were with native grass cover in the central quadrat,[5] with old remnants having higher cover than young, and soil pH in the central quadrat,[6] with young remnants having higher pH than old. These relationships are more likely to relate to changes in the types of country cleared (see Chapter 1) than a direct effect of age.

Effects of geometry on within-remnant variation

More isolated remnants are exposed to a greater proportion of exotics in the seed rain from surrounding areas than remnants that are close to other remnants. It could therefore be expected that they would have a higher exotic component, and less difference between their centres and edges.

The location of the remnants in relation to other remnants was important in explaining variation in condition between their centre and their edges. The ratio of exotic to native richness in the centre was significantly affected by the distance to the nearest larger remnant[7] and the distance to the nearest remnant,[8] with the more isolated remnants having higher proportions of exotic species in their centres than the less isolated remnants. The more isolated remnants had greater proportions of exotics in the centre than the cold edge, while the less isolated remnants had greater proportions of exotics on the cold edge than the centre.[9] The most remote remnants from other larger remnants tended to also have a greater proportion of exotic species in their centres than at their edges.[9] The more remote remnants from the nearest other remnant had greater proportions of exotics in the centre than the warm edge, while the less isolated remnants had greater proportions of exotics on the warm edge than the centre.[9]

Therefore, there is some indication that isolation of remnants reduces the differences between the centre and the edge on the condition variables. These results suggest that landscape effects need to be considered in planning patterns of remnant retention or native vegetation restoration. The planting of vegetation corridors to connect isolated remnants to other remnants can do no harm to the vegetation of the remnants, but is unlikely to mitigate the isolation effect unless the corridors have a highly diverse native species composition, rather than consisting of a few species of trees and shrubs.

The area of remnants was not related to variation in condition between their centres and their edges. However, there were several relationships with remnant width and one with the perimeter/area ratio.[10] Wider remnants tended to have a greater proportion of exotic cover on the cold edge, a greater proportion of exotic species on the warm edge, and have a lesser difference between centre and edge in the proportion of exotic species, than the narrower remnants. The proportion of exotic cover on both edges was more similar to that in the centre as the remnants became more compact.[10] These results may relate to the tendency of the larger remnants to have development in their centres.

There has been a tendency to discount small remnants for conservation action because of the high proportion of the area of such remnants subject to edge effects and because of a concentration on birds and mammals, rather than plants. However, in the study area, in the range 2–200 ha, the centres of remnants generally retain much higher conservation values than on their edges, with the differences being greater for the smaller remnants. These results, combined with those for age, support the earlier conclusion that small remnants can retain viability for long time periods (Kirkpatrick and Gilfedder 1995). As small remnants often support populations of rare or threatened vascular plant species (Kirkpatrick and Gilfedder 1995), their retention can be as desirable as that of larger remnants.

Effect of matrix on differences between edges and centres

It could be expected that differences between the centres and edges of remnants might vary with the nature of the surrounding land use. The greatest contrast in such use in the 100 remnant data set is between urban remnants, surrounded by suburbia, and rural remnants, surrounded by improved pasture and cropland. However, any comparison needs to take into account the differences between rural and urban remnants in several characteristics: the incidence of grazing, with ungrazed remnants being concentrated in the cities; the use of fire as a management tool, with fire being used in almost all urban remnants, but a low proportion of rural ones; the mean daily minimum temperature of the coldest month, with the rural remnants experiencing the colder temperatures; precipitation in the driest quarter, with the rural remnants being drier; the percentage of the stand logged, with the rural remnants being more likely to have been logged; the prevailing geological substrate, with urban remnants occurring only on the rocks that form clay soils and sandstone, while the rural remnants also occurred on Quaternary deposits and Tertiary sands and gravels.[11]

The urban remnants tended to have a greater proportion of richness attributable to exotics in the centre than the cold edge, while the reverse pertained to the rural remnants. The urban remnants tended to have a greater proportion of exotic species in the centre than the warm edge, while the reverse pertained to the rural remnants. Both classes of remnant tended to have a greater proportion of exotic species on the warm than the cold edge, with this tendency being more pronounced in the urban remnants than the rural ones.[11] These differences probably relate to the larger numbers of bird-dispersed exotic invaders that occur in domestic gardens than occur in improved pasture and cropland. The many frugivorous native and exotic bird species that are known to frequent Tasmanian gardens (Daniels and Kirkpatrick 2006) can disperse seed up to 400 m from the point of ingestion (Zacharek 1990). The increasing tendency for the establishment of residences and gardens in areas predominantly used for pastoral activity has obvious implications for the future of vegetation remnants.

Effects of management on differences between edges and centres

Stand management had some effects on the condition of the vegetation.[12] Both the cold and warm edges had a much greater proportion of exotic species in their floras compared to the centre as the time since the last fire increased. Heavy stock grazing was associated with high proportions of exotic species on the cold edges compared to the centres of the remnants.

Effects of environment on differences between edges and centres

There were no significant relationships between any of the condition ratios and any of the climatic variables. The cold edge quadrats on Tertiary sands and gravels had lower proportions of exotic species than those on Quaternary sands.[13] The central quadrats had greater proportions of exotic species on rocks forming clay-rich soils than on Tertiary sands and gravels.[13] Both of these relationships probably reflect the known favouring of exotic species richness by high fertility. Exotic cover is also thought to be favoured by high fertility. In support of this hypothesis, there was a strong effect of P, K and N content of the surface soil on the relative proportions of cover constituted by exotics between the centre and the cold edge.[13] A similar strong influence of soil macro-nutrients on vegetation condition has been found for roadsides (Cale and Hobbs 1991), remnant edges (Hester and Hobbs 1991) and suburban reserves (Clements 1983).

Remnant characteristics and management of biota

Different biotic groups could be expected to respond differently to the characteristics and management of vegetation remnants. Analyses have been undertaken of the responses of vascular plants (Kirkpatrick and Gilfedder 1995; Gilfedder and Kirk-patrick 1998a), birds (MacDonald and Kirkpatrick 2003) and bryophytes (mosses and liverworts; Pharo *et. al.* 2005) in subsets or all of the 100 subhumid Tasmanian remnants. These reveal a complex picture.

The results of the analyses of Pharo *et al.* (2005) show that the species composition of both bryophytes and vascular plants is influenced by geology and the mean daily minimum temperature of the coldest month. There the resemblance ends.

Bryophyte species composition, unlike that of vascular plants, is influenced by the geometry of the remnants, both distance to the nearest remnant and the perimeter/area ratio being present in models. The distance to the nearest remnant was also important in explaining bryophyte species richness. The spores of most bryophyte species travel in the wind. However, they are variable in this ability. Therefore variation in distance from a source region is likely to be related to variability in the species most likely to colonise, with a smaller subset being available the greater the distance.

The remaining variables that contribute to an explanation of bryophyte species composition are soil depth and pH in the edge quadrats, and mean annual precipitation. There are few bryophyte species adapted to survive in dry places, and many of them occupy rocks, which are most prevalent where soils are shallow.

Other environmental variables important in explaining vascular plant species composition were pH in the centre, the N/P ratio for soils on the edge and the mean daily maximum temperature of the warmest month. Vascular plant species composition also responded strongly to stocking rate, probably reflecting the fact that vascular plants are eaten by livestock, whereas bryophytes are not. The palatability of different plant species for sheep and cattle varies enormously (see Chapter 5).

The strongest influence on the distribution of individual bird species in remnants was the presence or absence of noisy miner breeding colonies. These tended to be in remnants that were less than 20 ha, had a high perimeter/area ratio, lacked a dense understorey and had much dieback (see Chapter 4). However, the spatial patterning of remnants was also important. Large areas of forest within 2.5 km of the remnant increased the likelihood of finding the common bronzewing, the green rosella and the sulphur-crested cockatoo. Long distances to the nearest forest patch greater than 50 ha reduced the likelihood of observing the grey fantail and forest raven, but increased the likelihood of finding the superb fairy wren.

Importance of threatened species in remnant management

While taxonomic groups have different patterns of response to remnant attributes and management, there is considerable variation in response between the species that constitute each of them. If conservation of biodiversity is accepted as the major function of remnant protection and management, the prime task is to preserve and manage for the more threatened species within any one remnant, and to preserve all remnants that support such species. This is because the survival of the community level of biodiversity depends on the survival of the species that constitute the communities, and because the surest way to reduce the species level of biodiversity is to allow the most threatened species to become extinct (Kirkpatrick 1999). As Kirkpatrick and Gilfedder (1995) have documented, threatened species in the Tasmanian subhumid remnants are not necessarily found in the largest remnants, or in those in best condition, partly because the disturbance regimes that allow some species to survive also promote the abundance and richness of exotics.

Notes

1 These images comprised Landsat 7 and Landsat 5 false colour satellite images covering the whole of the state, printed as colour composites in bands 2, 3 and 4, and ortho-corrected to AMG coordinates at a map scale of 1:250 000 (produced by the Australian Centre for Remote Sensing). The spectral bands selected were those that provided the greatest discrimination and ease of interpretation between areas of native vegetation, land that had been recently cleared and land modified by people for agricultural and other uses. The boundary between cleared land and native vegetation was mapped for the whole of Tasmania using 1988, 1994 and 1999 satellite images. This boundary was overlaid on satellite images at the same scale from the beginning of each of the two periods (1988–1994 and 1994–1999) and the areas of native vegetation cleared in the period determined. The satellite images could not be easily used to interpret all areas where natural vegetation had been lost to *Pinus radiata* or eucalypt plantations, because the early stages of plantation growth give the same signal as the early stages of native forest regeneration after clear-felling. Logging is not considered to constitute vegetation loss as the vegetation that regenerates is largely native. In the case of private land, field checking was undertaken to determine whether areas were plantation, agricultural land or regenerating native forest. A plantation map, dated 2000, was obtained from Forestry Tasmania and proved useful in discriminating some areas within public land. However, the totals obtained from this analysis were well short of the totals obtained from figures in the Forestry Tasmania Annual Reports, indicating that the map was not up to date. Therefore, the annual report figures for new plantation establishment in State Forest were used to calculate all totals. Each individual area of land cleared during the period was mapped and areas converted to plantation (eucalypt or pine), agricultural land or inundated by water, were calculated. The land clearance maps were overlain on a map showing the distribution of vegetation types (Kirkpatrick and Dickinson 1984). The areas of vegetation types cleared during the five-year period and converted to plantation or agricultural land were calculated and summed. The public forest plantation totals for each vegetation type were adjusted by multiplying by actual plantation area/mapped plantation area. The mean annual percentage loss of each vegetation type was calculated using the existing area figures in Kirkpatrick *et al.* (1995) as the base.

2 The data are from Fensham (1985). Extractable phosphorus is expressed in μg/g dry soil, using the method of Colwell (1965).

3 If a remnant still existed, and permission was given to access it, the following data were obtained: ownership; reduction in area; current management regime; cover of life form classes, rock, litter and bare ground in 10 m line transects located on the warm edge (north-west), the cool edge (south-east) and the centre of the remnant; a Bitterlich wedge count in the centre of the remnant. The location of the line transects followed the methods used in the earlier study. However, as the original line transects were not permanently marked, it is unlikely that any were in exactly the same position as in 1993. The 1993 and 1999 data for each cover class were compared using the paired t-test. The ownership data were placed in three classes: private, no change; private, change; public. The information on management and depletion was used to place each remnant for which permission was gained for access into five classes: remnant cleared; remnant reduced in area; management regime in 1999 less likely to maintain conservation values than management in 1993; management constant; management in 1999 more likely to maintain conservation values than management in 1993. Cross tabulation and Chi-squared were used to assess the relationships between the ownership and change in state variables. One-way ANOVA or Chi-squared was used to test whether there were any significant relationships, at $P < 0.05$, between simplified change in state classes (1 = cleared, depleted or poorer management; 2 = constant management; 3 = improved management) and each of the remnant attributes recorded in 1993.

4 The 100 remnants were in the range 5–200 ha in size, received less than 700 mm mean annual rainfall and were less than 500 m above sea level, making them all in the subhumid temperate climatic zone. Remnants that were visibly extremely degraded (i.e. with a high percentage of exotic cover throughout) were not

selected for survey. The age of the remnant (i.e. the time that had elapsed since it became isolated from a larger fragment of bush) was allocated to one of three classes (< 10 years, 10–40 y, > 40 years) based on aerial photography and verbal accounts from landowners. The following geometric data were collected for each remnant using maps and aerial photographs: area (ha); perimeter/area ratio (m/ha); distance to the nearest patch of bush (< 0.5 km, 0.5–2 km, > 2 km); distance to the nearest bigger patch of bush (< 0.5 km, 0.5–2 km, > 2 km), north-west–south-east width through the centre (< 300 m, 300–600 m, > 600 m). The surrounding matrix of the remnant was designated as rural or urban. Surface geology was placed in four classes: Quaternary sands; Tertiary sands and gravels; Triassic sandstones; argillaceous rocks (mudstone, dolerite and basalt). Climatic data were derived for each remnant using the Bioclimatic Prediction System (Busby 1988). The derived climate values we selected for use in the present study were: the mean daily minimum mean temperature of the coolest month; the mean daily maximum temperature of the warmest month; the mean annual rainfall; the mean rainfall in the driest quarter. At each site the presence or absence of accelerated erosion was noted. Samples of the A horizon were collected from the warmest edge and the middle of the remnant. Soils were stored in impermeable plastic bags and were later air-dried. They were then oven-dried for 24 hours at 105°C, ground and sieved. Total percentage nitrogen was determined using the Kjeldahl method (Jackson 1958). Soil pH was determined using the 1:5 soil/water suspension method (Rayment and Higginson 1992). Conductivity was calculated using a 1:5 soil/water extraction method (Rayment and Higginson 1992), and available phosphorus and potassium levels were determined using the bicarbonate extractable method (Rayment and Higginson 1992). The grazing history of the remnant was obtained from interviews with landowners or managers. Information was obtained on the stocking rate and whether the site was spelled regularly. Information was also collected on whether or not fire was used as a management tool and the time elapsed since the last fire. Transects 10 m long were placed, usually at right angles, to a north-west–south-east line passing through the centre of each remnant. The number of transects varied according to the length of this line. Three line transects were placed in narrow remnants (< 300 m average width in a north-west–south-east transect through the centre), five transects were placed in remnants 300–600 m, and for sites measuring > 600 m width nine transects were placed. Edge transects were placed one metre from the warm edge and the cool edge. If necessary, their orientation was adjusted to fit the edge of the remnant. A transect was placed in the middle of the remnant, and remaining transects were placed at equal intervals in between the centre and edge transects. Data were recorded from 437 transects. For each transect the presence of native and exotic higher plant species was recorded for an area of 10 x 1 m. The overlapping cover of native grasses, exotic grasses, native herbs, exotic herbs, native shrubs, exotic shrubs, lichens and mosses was measured along a 10 m tape, along with the amount of bare ground, litter and rock cover. Ratios of exotic species cover/native species cover (E/N.cov) and exotic species richness/native species richness (E/N.r) were calculated for each of the centre, warm edge and cold edge transects. These ratios were used to calculate: C/Co.cov (the ratio between the centre and cold edge in the ratio between exotic cover and native cover); C/W.cov (the ratio between the centre and warm edge in the ratio between exotic cover and native cover); W/Co.cov (the ratio between the warm edge and the cold edge in the ratio between exotic cover and native cover); C/Co.r (the ratio between the centre and the cold edge in the ratio between exotic richness and native richness); C/W.r (the ratio between the centre and warm edge in the ratio between exotic richness and native richness); W/Co.r (the ratio between the warm edge and the cold edge in the ratio between exotic richness and native richness). Ratios were also calculated for soil variables between the centre and the warm edge. Differences between each combination of centre, warm edge and cold edge transects in E/N.c and E/N.r, and the cover components were tested using the two-tailed paired t-test. This test was also used to assess the significance of differences between the centre and the warm edge on soil characteristics. Spearman's rank order correlation coefficient was used to test for the significance of relationships between continuous variables, one-way ANOVA was used to test the significance of relationships between categorical variables and continuous variables, and Chi-squared was used to test the significance of relationships between categorical variables.

5 ANOVA, $F = 4.9$, $P < 0.01$.

6 ANOVA, $F = 3.28$, $P < 0.05$.

7 ANOVA, $F = 3.49$, $P < 0.05$.

8 ANOVA, $F = 4.84$, $P < 0.05$.

9 The ratio between the centre and the cold edge in the ratio between exotic richness and native richness (C/Co.r) was significantly differentiated by distance to the nearest other remnant (ANOVA, $F = 4.25$,

$P < 0.05$). There was significant differentiation in the ratio between the centre and warm edge in the ratio between exotic richness and native richness (C/W.r) between classes for distance to the nearest bigger remnant, with the remnants that were most isolated from bigger remnants having higher values than those closer to such remnants (ANOVA, $F = 4.88$, $P < 0.001$). C/W.r was also significantly differentiated by distance to the nearest other remnant (ANOVA, $F = 8.47$, $P < 0.001$).

10 The north-west–south-east width classes were significantly related to: E/N.cov on the cold edge (ANOVA, $F = 3.54$, $P < 0.05$), with the wider remnants having higher values; E/N.r in the centre (ANOVA, $F = 3.86$, $P < 0.05$), with the wider remnants also having higher values; C/W.r (ANOVA, $F = 7.84$, $P < 0.001$), with the widest class of remnants having higher values than the two narrower classes. The perimeter/area ratio of remnants had a negative relationship with C/Co.cov (Spearman's $r = -0.216$, $P < 0.05$). Thus, the ratio of exotic/native cover on the cold edge was proportionately greater than the ratio of exotic/native cover in the centre as the remnants became less compact. W/Co.cov was also negatively related to the perimeter/area ratio (Spearman's $r = -0.267$, $P < 0.05$).

11 Rural and urban remnants differed in: the incidence of grazing (Chi-squared = 9.16, $P < 0.01$); the use of fire as a management tool (Chi-squared = 38.20, $P < 0.001$); the mean daily minimum temperature of the coldest month (ANOVA, $F = 22.81$, $P < 0.001$); precipitation in the driest quarter (ANOVA, $F = 14.84$, $P < 0.001$); the percentage of the stand logged (ANOVA, $F = 4.06$, $P < 0.05$); the prevailing geological substrate (Chi-squared = 17.60, $P < 0.001$). Urban remnants had significantly lower E/N.r in the cold edge quadrats than rural remnants (ANOVA, $F = 4.37$, $P < 005$; $F = 4.71$, $P < 0.05$). C/Co.r was significantly greater in urban than rural remnants (ANOVA, $F = 12.94$, $P < 0.001$). C/W.r was significantly greater in urban than rural remnants (ANOVA, $F = 6.67$, $P < 0.05$). W/Co.r was significantly greater in urban than rural remnants (ANOVA, $F = 9.00$, $P < 0.01$).

12 C/Co.r was significantly greater in remnants burned recently than in those not burned for a long time (ANOVA, $F = 4.94$, $P < 0.01$). C/W.r was significantly greater in remnants burned recently than in those not burned for a long time (ANOVA, $F = 3.37$, $P < 0.05$). W/Co.r was significantly greater in remnants burned recently than in those not burned for a long time (ANOVA, $F = 3.37$, $P < 0.05$). Heavily grazed remnants had lower C/Co.r values than ungrazed remnants (ANOVA, $F = 3.13$, $P < 0.05$).

13 The exotic/native richness ratio for the cold edge quadrat was significantly differentiated by surface geology classes (ANOVA, $F = 2.88$, $P < 0.05$). The same ratio was significantly differentiated by geology for the central quadrat (ANOVA, $F = 2.88$, $P < 0.05$). C/Co.cov increased significantly as the ratios between the centre and the edge in P (Spearman's $r = 0.432$, $P < 0.05$), K (Spearman's $r = 0.237$, $P < 0.05$) and N (Spearman's $r = 0.310$, $P < 0.01$) increased. W/Co.r was negatively correlated with the ratio between the centre and edge in soil surface pH (Spearman's $r = -0.353$, $P < 0.01$).

Chapter 7

The future of the run country

J.B. Kirkpatrick, A. Jensen and K.L. Bridle

> 'We will increase our return by improving the quality of our wool. The only way to make money is to grow the best to make the best.'
>
> 'Strange, I'm starting to get more green.'

Summary

The interaction of wool growers with nature conservation bureaucrats has not always been happy, especially in relation to wildlife management. While clearance of conservation-significant forest vegetation has been controlled on the runs, highly significant non-forest native vegetation remnants and wetlands are still being cleared, drained and cultivated, largely in response to recent inept political moves to control such clearance. Graziers distrust the ability of government to understand the exigencies of their operations, and fear that they will be forced to subsidise the conservation costs of society in general. Realistic payment for ongoing conservation services combined with sympathetic regulation appears to be a feasible solution. As shown here, such payments would not be so large as to render them impractical. Assuming no reversal of the ongoing decline in wool prices, wool-producing sheep may only continue to run in native pastures if premium markets are enticed through good story-lines based on accreditation of outstanding environmental practice. The potential for a good marketing story, based on reality, is certainly there. The mechanism of biodiversity plans integrated with property plans may help in both accreditation and nature conservation.

Introduction

The first decade of the twenty-first century is a difficult time for wool producers in Australia. Wool prices are low (see Chapter 1, Figure 1.22) and seem unlikely to improve; competition from China is increasing at the bottom end of the market; the work of the animal rights movement increasingly threatens some markets; and there is an increasing tendency of governments to regulate on-farm activities for environmental and social outcomes. These outcomes do not necessarily enhance the bottom line of an enterprise.

The decade has also proved difficult for those who wish to conserve nature. The focus of government conservation efforts on old-growth forests, in response to sustained public pressure, has left little effort and resources to devote to the species and communities that are most threatened by human activity. Nature conservation bureaucracies have been atomised, submerged and deskilled. Government actions to solve biodiversity conservation problems have been partial and only partly effective, often creating hostile responses. However, within any set of problems there may lie opportunities.

A thoughtful and well-informed wool grower described some possible future scenarios for the run country:

> 'The current predominant use (of runs for) grazing by wethers for wool production is very much dependent on markets. At the current level it is not feasible to increase grazing intensity through intensive grazing regimes, or other management options that involve more inputs – labour or cash! Most properties are concentrating on their arable areas and cropping/irrigating these.
>
> Two scenarios, this stays the same or gets relatively worse, then the level of grazing will decrease further, or if price increases, then more attention will be paid to wether management for wool, with increased stocking rates and management inputs I suspect. The related issues of fencing for more effective game management and burning are also very much tied to the wool market. Burning is becoming increasingly difficult due to restrictions and time required and, given the public attitude to risk, I suspect the regulations will make wide-scale burning pretty much impossible.
>
> Recreational hunting can also go two ways. If greater gun control measures are implemented, culling of wallabies will be more difficult. I suspect there will be plenty of deer hunters around! Alternatively, hunting may become a lucrative activity for landowners, with restricted access to people paying for the privilege!
>
> There are some (conservation) programs around currently that could supplement grazing. At the current levels of payment ($200/ha for covenant) they are not very attractive, but if they were management agreements for a limited time they could be taken up.
>
> There are related tourism/recreation opportunities, focusing on natural values, but they are likely to be restricted to areas of outstanding beauty, with fishing/hunting opportunities. For most landowners there is a strong

> sense of attachment with the land they manage, and this is particularly so for natural areas. I am not sure how this will unfold in the future or what the consequences might be, but I suspect this will be ongoing and mean that landowners will want to retain these areas and get whatever return they can, without affecting their own values for the areas. I think it unlikely that these areas will be sold off to any extent, although what areas do change hands are likely to be managed primarily for uses other than grazing.
>
> The other issue is climate change. There has certainly been a drier trend over the last 10–20 years. I guess this means less trees, particularly on shallow soils – white gum areas, where tree growth is certainly constrained by moisture. For brown peppermint I am not sure. Is moisture a driver for these communities or is it something else? I guess my parting comment would be that a strong wool market will ensure a continuation of the current management, most of which is pretty good in terms of retaining the diversity etc. Reduced grazing will mean more trees/shrubs.'

Rather than developing scenarios, this chapter focuses on opportunities to integrate profitable wool production with nature conservation on the run country of Tasmania. The recent history of interaction between graziers and nature conservation bureaucrats is described, attributes of an effective, fair and acceptable means of conserving nature on private land are discussed, the degree of potential conflict between nature conservation and production is quantified, and possibilities for using nature to help the economics of wool-growing enterprises are mooted.

Bureaucrats and graziers

In the early twentieth century the State of Tasmania adopted the idea of the national park, an area set aside for the preservation of nature and the recreational enjoyment of locals and tourists. Mount Field National Park and Freycinet National Park were among the first in Australia (Mendel 1999). Until the early 1970s, these parks were selected on the basis of their scenic grandeur and their uselessness for primary production (Mendel and Kirkpatrick 1999). After 1970, they became increasingly chosen for their wilderness value (Mendel 2002) or their representation of biodiversity (Mendel and Kirkpatrick 2002). Some of the national parks and other reserves declared in the massive expansion of protected areas in Tasmania between 1972 and 2005 were used as runs for sheep, most notably parts of the Maria Island National Park, the western section of the Schouten Island addition to Freycinet National Park and much of the eastern Central Plateau addition to the Tasmanian Wilderness World Heritage Area. Sheep were eventually removed from all the above areas (see Chapter 1 for the Central Plateau story). These removals were not liked by most graziers:

> 'Maria Island a disgrace, took farmers out, a mess.' (Gilfedder and Kirkpatrick 1995a, p. 68)

Peter Murrell, the first Director of the Tasmanian National Parks and Wildlife Service, which was set up in 1971, believed that nature conservation could only be effectively achieved by reservation under the Act that he administered (personal communication to J.B. Kirkpatrick, 1975). He had a particular fondness for reserving islands, and was not averse to compulsory purchase, as in the case of Partridge Island in 1975. However, almost all of the area of reserves set up under his directorate was on public land, very little of which had any past agricultural or pastoral use, or potential for such use.

In the 1970s the major interaction of the Parks and Wildlife Service with graziers was in its role in regulating and policing the taking of wildlife. There was some very positive cooperation between farmers and the Service. For example, in efforts to ensure the future of the forester kangaroo, several farmers volunteered to host populations on their properties, on which the species had previously become extinct. However, control of wildlife management by the service was not regarded positively by many in the wool-growing community, as indicated by some of the responses in the interviews of Gilfedder and Kirkpatrick (1995a):

> '... much suspicion of Parks and Wildlife in community, especially wildlife issues.'
>
> '(Parks and Wildlife) not trusted by a lot of farmers.'

Suspicion and lack of trust was, to some degree, mutual. Employees of the Parks and Wildlife Service saw their role to be to ensure the survival of native species in the landscape, not to ensure the economic viability of wool producers, while wool producers perceived no likelihood that the mammalian pests of their pastures would become extinct on their properties, as much as some would have valued this outcome, and were focused on making a living from the land. They did not appreciate any bureaucratic control of their activities, and resented the time spent in gaining bureaucratic approvals.

Debates over hydro-electric development, forestry and wilderness, in which the Service argued for conservation, made development-minded governments increasingly suspicious that it was the bureaucratic arm of the Greens. In likely consequence, the Parks and Wildlife Service was submerged in the Lands Department in 1993, and then sunk deeper, with much of the Lands Department, in the Department of Primary Industries, Water and Environment (DPIWE) in 1998. Parks and Wildlife became much more under political control, as upper level bureaucratic positions were converted into short-term contracts.

The transfer to DPIWE was made with a government expectation that the primary industries focus of the Department as a whole would moderate the conservation ethic of Parks and Wildlife. In 1996, Max Kitchell, a Victorian who developed programs for nature conservation on private land in his home state, was appointed as Manager of the Parks and Wildlife Service. He encouraged the development of programs to involve the rural community in nature conservation, particularly Land for Wildlife. The voluntary nature of this program was attractive, as was the lack of any reservation security for land marked with attractive green signs, and the potential to gain useful management information:

> 'Voluntary scheme OK because farmer still has control.'
>
> 'Yes, keen to be involved in a voluntary scheme.'
>
> '... get help with species information, management, better ways.' (Gilfedder and Kirkpatrick 1995a, p. 65)

However, many farmers were reluctant to participate because of lack of interest, lack of time, or because they saw nature conservation on their land as their family responsibility:

> 'Land for Wildlife has its appeal, another time-consuming activity.'
>
> 'Voluntary schemes – not with government, best a family thing, don't want to get involved.' (Gilfedder and Kirkpatrick 1995a, p. 65)

By the late 1990s, there was an increasing tendency to consult and cooperate, rather than order and control, within Parks and Wildlife. Even so, the scientists and other professionals employed by Parks and Wildlife were still sometimes effective in preventing or moderating development proposals within the reserve estate whenever they judged that they would have a negative effect on natural or cultural values. However, the State Government at this time wanted private tourism development within parks, and wanted a Parks and Wildlife Service that gave precedence to servicing the tourism industry, not to conserving the natural and cultural values of the parks. The solution was to split off the biologists, geomorphologists, historians and archaeologists, leaving the park managers, planners, interpreters and fire managers under the control of a new manager who reworked both mission and attitude more to the taste of government. In 2002, the pruned Parks and Wildlife branch was transferred to a department that had tourism as its major responsibility, and thereafter had no major conservation or management role on private land.

In 1995 the Tasmanian Parliament approved the *Threatened Species Protection Act* (TSPA), after taking advice from a committee largely composed of rural landowners and biological scientists. This Act set up a list of species thought to be threatened at the time, a Scientific Committee with the responsibility to provide advice to the Minister on the listing and de-listing of species, and other scientific matters, and a Community Review Committee which had oversight of the social and economic consequences of the Act, and could advise the Minister on these matters. A permit was required to take listed species or to destroy the critical habitat of listed species. While the Minister could halt an action that threatened the critical habitat of a listed species, this action was not compulsory, and compensation was ultimately required. The TSPA has only inconvenienced wool producers who wish to build large dams or other major constructions. The development approvals process for such major developments requires a determination of possible impacts on listed species, which, in turn, usually requires an expert survey, and an application for a permit to take individuals, if listed species are located, and if the development cannot be modified to avoid destroying them. Applications for such permits are seldom refused.

Clearance or modification of native vegetation with listed species or their critical habitat could theoretically be prevented, or punished, under the TSPA. However, the Threatened Species Unit of the Department of Primary Industries and Water, which administers the Act, has been reluctant to take legal action, largely because it judges that it is likely to be ineffective because of the contents of the Act. Instead, it has taken an educational approach. Given the nature of the Act, it is not surprising that some graziers prefer not to know if threatened species occur on their runs, and prevent access by biologists. Other graziers delight in their threatened species, and take good care of them (see Chapter 3). Threatened species are, by definition, not very common. If they do prove to be common, they are de-listed. On most wool-growing properties they therefore have very little potential to disrupt normal land use, even if they were fully protected.

The signing of the Regional Forest Agreement (RFA) by Tasmania and the Commonwealth in 1997 drew the conservation importance of the bush country on the runs into close focus. The *Forest Practices Act* of 1985 had imposed obligations to develop forest harvesting plans and to conform to the Forest Practices Code, if private forest was cut beyond a minimal level. There was no control of land clearing in this process. Many graziers managed bush runs to provide an income stream from wood sales. Others used the sale of wood to subsidise conversion of bush into improved pasture.

The RFA contained clauses that required the protection of threatened forest species, and rare or threatened forest types, and other clauses that limited the total amount of native forest loss by region. These clauses caused the Forest Practices Board to modify their approvals system in order to limit private rights to clear forest. Unlike the example of protection of threatened species under the TSPA, these limitations were, and are, policed and enforced. The changes proved generally acceptable to rural landowners, probably because the Forest Practices Board was seen as a body sympathetic to production interests, and because the new administrative processes caused little change in their bureaucratic burden or the potential to sell wood. In cases in which harvesting or clearing were prevented, some financial compensation was available through the mechanism of the Private Forest Reserves Program (PFRP), funded under the RFA (see Chapter 3). However, a permanent covenant was required to access these funds.

In 1999 the Parliament of Australia approved the *Environmental Protection and Biodiversity Conservation Act* (EPBC), which gave the Government of Australia the power to protect nationally threatened species and communities listed under the Act. Communities had not been protected under the Tasmanian Act. Under the EPBC, any development that could potentially affect listed species or communities, and was not a matter of normal rural land use, or covered by the RFA, had to be referred by the proponent to the Australian Government to make a decision on the need for further assessment. Ultimately, the responsible Australian Government minister could prevent a development that they judged to be a threat to the future of listed species or communities.

Many species that were listed under the Tasmanian Act were not listed in the EPBC, because, while threatened in Tasmania, they were safe on the mainland. The threatened non-forest plant communities that occur on the runs of Tasmania were not listed in 1999. However, there is a current (2006) process to assess whether Tasmanian native grasslands should be listed, and how the grasslands as a whole, and the grasslands of highest quality, should be defined. The listing of native grasslands under the EPBC may, or may not (ministerial decisions are involved), restrict clearance of the highest quality remnants. Judging from the experience in other states where grassland has been listed under the EPBC, it will not interfere with pre-existing use for grazing, and will not prevent most remnants, which are of moderate or low conservation quality, from being cleared. From a conservation perspective, high quality native grasslands are exceedingly rare.

The sale of half of Telstra, the Australian Government-owned telecommunications company, was pushed through the Senate by one vote in 1996 after a deal was struck with the Tasmanian Senator Harradine. This deal involved a disproportionate amount of the money from the Natural Heritage Trust (NHT), which was funded from the sale of Telstra, being directed towards his state. A substantial amount of the Tasmanian NHT money was spent on wool-growing properties through community organisations such as Landcare, Rivercare, Coastcare and Greening Australia. It was used to subsidise fencing of bush and rivers, for tree planting, weed control and erosion control, and for employing large numbers of extension officers. Where the money was spent on fencing native vegetation, there was no requirement for permanent protection: 'Newly developed ground fenced off with Landcare seven to eight years ago.'[1]

By the end of disbursement of the first sum of money dedicated to the NHT, there was general governmental dissatisfaction with a large proportion of the projects that were funded. They were regarded as scattergun, rather than strategic, in their incidence, and ephemeral in their impact. The Australian Government was suspicious that much of the NHT money was used to perform conservation functions that would otherwise have been supported by the state. In the case of Tasmania, the Australian Government was also upset that the State Government had not delivered on specific obligations that it accepted as part of the agreement to provide NHT funds. One of these obligations was to set up a system to control land clearance where it threatened the conservation of native biodiversity.

The second round of NHT did not favour Tasmania in its disbursement of funds, Senator Harradine's vote no longer being necessary. Bureaucratic solutions to perceived problems seldom reduce the size or complexity of governance arrangements. The somewhat baroque solution for the problem of non-strategic use of NHT funds was the expensive setting up of regional entities called Natural Resource Management (NRM) Boards which had the tasks of developing NRM strategies for their region and assembling and disbursing funds to achieve their strategic goals. An Act of the Tasmanian Parliament set up three of these boards in 2002, appropriately using the Telstra phone book regions, and prescribed the mixture of expertise of those who were to compose them. The positions were advertised, and members and chairs selected by the relevant minister.

Strategy development is always a protracted process in these days of inclusiveness. The development of NRM strategies for the three regions required massive consultation of a public who had barely emerged from being consulted to death in the Tasmania Together statewide vision development process. The period between NHT1 and the approval of regional strategies by the Australian Government saw the various 'care' movements, and organisations like Greening Australia, slide into penury. Most of the horde of extension officers and facilitators, who were just coming to grips with the practicalities of their jobs, found themselves unemployed. The communities of voluntary 'care' workers had no special connection to the NRM Boards and no access to help or money. They were not happy, with many losing the heart to continue their work. A small project funding scheme, 'Envirofund', independent of the NRM Boards, was eventually created to try to hold the volunteers.

The funding to the state for NHT2, under the agreement between the Tasmanian Government and the Australian Government, was made contingent, among other things, on the Tasmanian Government initiating and implementing a process to control land clearance where it threatened conservation-significant non-forest vegetation. An act to protect threatened communities was passed by the Tasmanian Parliament in 2006. It excluded lowland native grassland and grassy woodland from a list that contained much less threatened communities.

The protracted period of consultation, without any interim control or compensation measures, was guaranteed to raise the rate of clearance:

> 'There's a mad push to plough everything that's ploughable (due to the prospective non-forest clearance legislation). Most people are ploughing stuff that was partly done years ago. Legislation drives people to extremes. They will do it in case they get stopped from doing it. They'll do it straight away. If they had the right, they wouldn't be bothered.'

In a predictable conservation tragedy induced by this appalling series of political decisions, part of the best lowland grassland remnant in the state, containing many listed species, was ploughed up in 2006. Graziers have also wasted their money and time ploughing up much semi-natural vegetation of negligible conservation and production significance.

Government supervision of the management of native grassland and grassy woodland, the very basis of economic production in the runs, is seen by wool producers to be much more hazardous to their enterprises and well-being than limitations on forest clearance. Some of those graziers who have been most active in conserving nature on their properties are among the most aggrieved at the prospect of legislative controls on non-forest vegetation. They have seen the management agreements attached to covenants under the PFRP program and have not necessarily been impressed with their likely effectiveness in conservation, much less their awareness of the needs of overall property management. They know that the parts of the agencies that would manage any process are understaffed, that their staff who have experience in property management are few, and that staff turnover is high:

> 'As a wool grower there is nothing more frustrating than a constant turnover of people in these positions. New relationships need to be built and the incoming personnel need to be brought up to speed – a waste of wool growers' precious time.'

They fear a formulaic, legalistic imposition of management regimes that might work in one place, but not in another (see Chapter 2). The 'another' might be theirs. Above all, they feel that their efforts and sacrifices for nature conservation on their properties are more likely to be punished by any process than rewarded, and their sense of doing the right thing for their own reasons will be violated:

> 'We treasure what we've got and not having someone to tell us how to run it.'

> 'Those who have done the right thing appear to now be penalised for doing so!'

All graziers who talked on the subject thought that there should be a limit on the degree to which property owners were expected to provide for the social and environmental desires of the community as a whole:

> 'Compensation – any financial advantage would be an incentive – we're providing a public good with conservation and they (the public) should contribute.'

It may well be that, when a system eventually emerges to protect the little native grassland and grassy woodland that will be left, the types of problems anticipated by graziers will be overcome. In the next section we discuss potential attributes of a solution that could have graziers happily managing for nature conservation and nature conservationists happy with this management.

A nature conservation system

There have been many different approaches used to help achieve the goal of nature conservation on private land (Binning 1997). All of the major approaches contribute to the achievement of the goal, while being insufficient to achieve it alone.

Landcare, Bushcare and the other 'care' movements have been, and are, highly effective in promulgating awareness of the need for sustainable use, and in spreading knowledge of the management techniques that could implement it (Campbell 1994; Curtis 1997). As a result of the activities of these movements, management has become more sustainable on a significant proportion of Australian rural properties. However, a substantial proportion of landowners have not been affected in either their ideas or practices, many remaining focused on maximising short-term returns from their land and being unconvinced of the virtues of nature conservation, at least on their own properties (Gilfedder and Kirkpatrick 1995a). It seems very unlikely that the 'care' ethic will ever achieve total penetration. Without other mechanisms, property turnover has ensured, and will ensure, continued degradation (see Chapter 6).

Voluntary commitments and covenants are another widely used mechanism. These vary from the 'Land for Wildlife' type of scheme where owners make a non-legal commitment to keep bush for conservation management, to upfront payments of varying magnitude to maintain bush in perpetuity through a covenant on the title. Non-legal commitments and short-term legal agreements cannot achieve stewardship in the long term, because of property turnover (see Chapter 6). Many of the perpetual conservation covenants have been signed by people who live in the cities and have purchased land in the countryside for recreational and conservation purposes. This has resulted in a bias towards near-coastal and forest land within two hours' drive of the cities. In these areas covenants increase the value of most properties. Further from the cities, and in less attractive situations, perpetual conservation covenants may lower the value of properties, with payments usually insufficient to cover these losses, not to mention the cost of the legal obligation to manage the covenanted bush for conservation purposes, forever. In north-eastern Victoria, properties consisting of more than 50 per cent protected remnant native vegetation gained a lower sales price than those that did not. Below 50 per cent there was no effect on prices (Walpole *et al.* 1998; Lockwood *et al.* 2000). The effectiveness of covenants depends on enforcement after changes in ownership. The mechanisms for such enforcement appear to be weak to non-existent. With negligible enforcement from government, covenants are likely to be effective in retaining bush in the long term only in areas where members of the community support retention to the degree to which they are prepared to force government action. Such public pressure is likely to be ineffective in ensuring appropriate management of covenanted vegetation. This depends on the goodwill of the new landowner.

Regulations controlling land clearance and land management vary in their effectiveness, depending on the degree of their acceptance in the places to which they have been applied. Laws and regulations that are not believed in by most of those to whom they apply are seldom effective. Even under the most stringent regulations in Australia today, clearance by attrition or stealth is legally possible. Small areas can be cleared without the regulations applying, and destruction of bush by neglect or mismanagement is almost impossible to regulate. In most jurisdictions, criteria are applied to assess whether larger clearances are permissible. These raise the possibility of legal argument about the meanings of particular words and phrases.

Payment for service is another option that is likely to be ineffective by itself. In the European Union, farmers are paid to maintain archaic land management regimes that are important for biodiversity maintenance (Kleijn and Sutherland 2003), while in the United States, farmers are paid to not produce crops, and thereby conserve soils (Binning 1997). At least, those are the arguments. Some interpret the payments as unfair subsidies to the farming sector, because the method of payment is not the most economically efficient way to achieve the desired outcomes.

Payments for biodiversity maintenance have not worked when either party could withdraw on short notice, as opportunity costs, and government funding, vary through time and renegotiations are expensive and time-consuming. Continuity of management regimes is critical in effective nature conservation.

Open and voluntary eligibility for payment is an inefficient way of achieving nature conservation, as there is no assurance that the money will go to where it could be best used. For example, payments for not clearing are likely to be more attractive to people with native vegetation on less productive soils, that have vegetation types that are unlikely to be threatened.

If payment is to be used effectively in a nature conservation system: those to whom it is applied, and who apply it, should perceive it to be beneficial to them; rewards need to reinforce performance; the outcomes should be permanent; and, the system should be administratively and financially efficient.

Farmers are used to providing goods and services for recompense. They are generally highly positive about the idea of payment for nature conservation services, as long as the payment is 'realistic' in relation to the costs involved, including forgone options to make money (opportunity costs). Tax deductions do not work as well as direct payments in situations in which a large proportion of rural enterprises may have low or negative net incomes in many years. Compensation payments do not provide an incentive for continuing management.

To be effective in achieving its goals a nature conservation system should reward their achievement and penalise their non-achievement. For example, if payments are received for the service of maintaining a population of a threatened plant species on a property, and that population dies out, payment should cease.

Nature conservation systems cannot be effective if they are totally voluntary in time or space. Negative impacts on biodiversity either cannot be rectified, or take considerable expense and time to reverse. Permanence of outcomes requires an underlying level of regulation. It is important that such regulation focuses on the critical outcomes, for example: no net loss of native vegetation cover; maintenance or increase in the populations of threatened species; and, reduction of stream salinity and turbidity. Detailed regulatory prescription of methods to achieve such goals devalues the knowledge and experience of the manager, and is unlikely to be universally effective in nature conservation.

The most efficient systems are those that have clear rules and procedures and auditing to the degree necessary to ensure general compliance. Efficient systems also have strong mechanisms for directing scarce funds to the highest priorities, and do not use these funds to pay for actions and activities that would occur without payment. It has been suggested that a 'duty of care' would require that every property owner keep at least 10 per cent of their property under native bush, and there is also much bush on private properties that would not repay clearance using current technologies. There is other bush, which, when covenanted and in good condition, adds to the value of properties. In short, payments for service need to be directed to where they will achieve outcomes.

In one of many possible permutations of an effective nature conservation system that satisfies the requirements discussed above, the landowner could apply for a stewardship payment agreement to a properly constituted body (see below). Such an application could require a property plan, in which the distributions of vegetation types, rare or threatened species and their habitats, and land capability classes were mapped in a GIS environment, as well as other assets of the property, and potential changes to land use. These data could then be overlain to determine the areas of conflict between regulatory requirements and potential production.

For example, there may be 50 ha of land use capability class 3 land that cannot be cleared because of regulations and 100 ha of class 5 land in the same situation. The payment could be the difference between the current net income under bush and the potential net income under the planned use, as long as the planned use was sustainable, given the land capability, and legal. Prevailing discount rates could be used to determine the potential costs and benefits in current dollars.

Returns and costs could be calculated over a decade to take into account necessary rotations and the cost of replenishment of nutrients. In the above calculation, costs could include the cost of clearance and the cost of transferring produce to market, as well as input and maintenance costs. This opportunity cost payment could be discounted 10 per cent for duty of care and could be subject to cost of living adjustments, but not to future recalculation on the basis of variation in the costs of inputs and prices of outputs for the alternative productive use.

Additional payment could be requested for the mitigation of threats to nature conservation values, such as weeds or feral animals, where these costs could be shown to be over and above the costs normally incurred in the farm enterprise, and only to the degree to which they could be shown not to directly benefit the production economics of the property. These payments could be renegotiable on the basis of new threats or major variation in the cost of management inputs. The agreement could have targets and monitoring procedures. Performance against targets could be self-reported and subject to random audit. The stewardship payment agreement could legally bind all future owners of the property, and could be varied only with the mutual consent of the owner and the regulating authority.

The minimum standard for assessment of applications could be acceptance by a committee with expertise in farm management in a region similar to that of the applicant, farm economics, land care, soil science and nature conservation management, operating within guidelines determined by intergovernmental agreement, and using data provided by the Australian and State governments. Negotiators could be employed to finalise agreements, and auditors employed to determine conformance. The NRM regional boards could be well placed to supervise this assessment and auditing.

The above system would need to sit within a regulatory framework. Some laws and regulations to do with private land are absolute: thou shalt not grow illegal plants; thou shalt not light a fire that burns into your neighbour's property. Most of the laws and regulations that cover the potential destruction of nature conservation values on private land are not absolute. Rather, they require the gaining of a permit to destroy, resulting from a judgement on the part of a bureaucrat, politician, judge or committee that such destruction does not violate particular thresholds or goals. For example, in some jurisdictions there are legal targets for the proportionate retention of native forest, which mean that permits can be given until the limit is reached, thereby postponing consequent conflict. Permits to take threatened species are commonly given because the numbers of individuals to be taken are small in relation to the species as a whole. It is hard to judge when the incremental effect of granting such permits presents a significant risk to the future of a species. Combining this fuzzy regulation with stewardship payments has the benefit that protection is successively built up at the same time that populations are successively destroyed. Monetary rewards for maintaining populations of threatened species or areas of threatened ecosystems are likely to lead to increases, rather than the decreases through neglect that could be expected if threatened biodiversity is seen as a handicap to economic development.

There has been no shortage of suggestions that our knowledge base needs to be dramatically improved before we can adequately conserve biodiversity on private land through a stewardship payment system. While there is no doubt that more research will be highly beneficial to such efforts, we risk substantial nature conservation losses if we do not proceed on the basis of current knowledge. We have regulatory frameworks that specifically recognise threatened species and communities, using the best scientific information available. A stewardship payment system could be based on these listings.

Much further research is also needed on the management requirements for biodiversity conservation in agricultural and pastoral landscapes. A stewardship payment system dependent on maintenance or improvement will encourage research directed towards this end. It is likely that appropriate management regimes for particular entities or qualities may be already occurring. After all, if there are threatened elements of biodiversity on a property, these elements must have survived previous management regimes. Nevertheless, it is prudent to experiment with small changes in small parts of a property, then change management more widely if the results of experiments indicate benefit, as the survival of a threatened species under a particular management regime does not necessarily mean that it is not on the way out.

The system described above could use already established structures and organisations in its administration. A substantial proportion of public and private conservation-related expenditure is currently devoted to nature conservation on private land. Most of this could be redirected into the administration and funding of a stewardship program. However, to be comprehensive and effective it seems likely that some further net transfer of funds from the urban regions to the rural regions will be necessary. An economic analysis might show no net cost, as a flow of funds to the countryside, inasmuch as it increases employment, may result in more efficient use of existing infrastructure, which has been underused as a result of rural depopulation, and reduce the need for more investment in the urban regions.

The size of the investment that would be required to run the system may not be very much larger than the existing investment. Rural private and leasehold properties that have nature conservation value on land that has potential for further development are likely to be a small proportion of the total number of properties, and, in most cases, very little of the property would be affected. Nevertheless, in some individual cases, the effect could be substantial. The following section attempts to quantify this effect for part of the Macquarie River catchment, an area highly important for both wool production and nature conservation in Tasmania.

Opportunity cost of nature conservation

The dependence of threatened species and communities on the perpetuation of a variety of agricultural management regimes is a commonly recognised phenomenon in Europe (see for example Muller 2002; Sutherland 2002). However, it has received little attention in the New World, although Freemark *et al.* (2002) in North America emphasised the need to maintain a diverse mosaic of habitats on farmland to conserve native plants, and Lunt (1997) in Australia demonstrated that a diversity of management regimes was necessary to maintain grassy ecosystem diversity.

In Europe it is common practice to subsidise archaic agricultural management regimes for the purpose of biodiversity conservation, often using a profits forgone (opportunity cost) approach (Kleijn and Sutherland 2003). In Australia, the emphasis has been on finding politically, socially and economically acceptable mechanisms to prevent the clearance of native vegetation on private land (Binning 1997). Where these approaches involve monetary incentives or compensation, they can value on the basis of the financial implications of exclusion of commercial use, costs associated with management, and the conservation value of the vegetation (Binning 1997). Yet, where, as in much of the run country of Tasmania (see Chapter 5), a change in land use management can be a threat to biodiversity, and current land use is not a threat, an opportunity cost approach seems desirable.

Since the initial work of Kirkpatrick (1983), a wide range of techniques has been developed to find reservation solutions that minimise the area set aside from other uses in order to satisfy specified nature conservation criteria (Margules and Pressey 2000). These techniques can equally be used to select a set of areas that minimise any other criterion, such as opportunity cost.

In Australia the knowledge of groups of experts has been utilised in the so-called 'Delphi technique' to determine both conservation targets at a property scale (McIntyre *et al.* 2002) and a bioregional scale (Kirkpatrick 1998). As the results of the application of such criteria are likely to be scale-dependent, there is a need to address the implications of these two scales of analysis on landscape patterns of native vegetation retention.

The Northern Midlands of Tasmania, part of Australia's second oldest agricultural area, is considered a threatened bioregion (Commonwealth of Australia 2001). Less than 30 per cent of the region remains as native vegetation, and over 180 species that occur in the region are listed under the *Threatened Species Protection Act 1995*. A number of plant communities, largely grasslands and grassy woodlands, have been largely eliminated. These include *Eucalyptus ovata – E. pauciflora* grassy woodland (restricted to 2.3 per cent of pre-European estimated range) and *Themeda* grassland (restricted to less than 12 per cent of pre-European estimated range) (Gouldthorpe and Gilfedder 2002). Over 95 per cent of the bioregion is under private ownership.

The 63 020 ha study area, used to determine the implications of vegetation and threatened species conservation on opportunity cost, consisted of nine properties and adjacent state forest and reserves in the upper Macquarie River catchment of the Northern Midlands.

The upper Macquarie River catchment contains one of the driest areas in Tasmania, located within the rain shadow of both easterly and westerly systems. Average annual rainfall ranges from 450 mm in this area to 800 mm on the adjacent ranges. Average daily maximum temperatures range from approximately 11°C (July) to 24°C (February). The dominant land uses are sheep and cattle grazing, dry land and irrigated cropping, and forestry.

Seventy-seven per cent of the study area was covered by native vegetation. Twelve per cent of the study area was in public or private reserves and 22 per cent in state forest dedicated to wood production. The remainder almost totally consisted of privately owned agricultural properties, which contained almost all of the cleared land and more than half of the native vegetation.

Thirty-nine native vegetation communities occur in the region. Five of the 39 native vegetation communities present in the study area are adequately reserved (> 15 per cent of area before clearance in reserves) at a statewide level, these being tall and medium height *E. delegatensis* forest, *E. obliqua* wet and dry forest, and *E. pulchella*/*E. globulus*/*E. viminalis* forest (Mendel and Kirkpatrick 2002). None of the woodland communities are adequately reserved at the statewide level. Reservation of natural grasslands is extremely poor, with the 16 ha Township Lagoon site at Tunbridge the largest secure site for natural *Themeda triandra* grassland in the state (Kirkpatrick *et al.* 1995; Zacharek *et al.* 1997).

Vegetation communities of very high significance for conservation in the study area are *E. ovata*/*E. viminalis* woodland, *E. ovata*/*E. pauciflora* grassy woodland, *E. pauciflora*/*E. viminalis* woodland, lowland *Poa* grassland, *Themeda* grassland, shrubby *E. ovata*/*E. viminalis* forest, *E. viminalis* wet forest on basalt, riparian vegetation and wetlands. These communities have all been restricted to less than 20 per cent of estimated pre-European range within the Northern Midlands, and are very poorly reserved, if at all (Kirkpatrick *et al.* 1995; Gouldthorpe and Gilfedder 2002). Saline or brackish wetlands are also considered to be of very high significance for conservation, due to the large numbers of threatened species they support.

Within the study area, the Threatened Species Unit (Department of Primary Industries, Water and Environment) had known locations for 48 plant species listed under the *Threatened Species Protection Act 1995 Tasmania* at the time of the study. Seven of these species are endemic to Tasmania.

Tasmania's agricultural land has been modelled for land capability, or suitability for agriculture, a function of soils, topography and climate (Noble 1992). Factors constraining agricultural use include shallow soil depth, coarse rock fragments, poor drainage, unreliable rainfall, erosion hazard and other hazards (such as climatic extremes or salinity). Scores range from 1 (the most versatile land, well suited to intensive agriculture) to 7 (scree slopes). Public lands have generally not been classified.

Classes 4, 5 and 6 are present in the study area. Class 4, covering 7 per cent of the study area, has the greatest potential for agricultural activities. This land is well suited to intensive grazing and some cropping, though it is suggested that cropping be restricted to one or two years in 10 (Grose and Moreton 1996). Class 5 land covers 25 per cent of the study area. Cropping is considered unsuitable on this land, but pasture improvement is possible on gentler slopes. There are slight limitations to pastoral use. Class 6 land covers 34 per cent of the study area. This land occurs on the slopes of the catchment, has a high risk of erosion and is unsuited to pasture improvement. It is considered marginally suited to grazing. The remainder of the area, all state forest, was not mapped, but is likely to be class 6 land. Modelled land capability is strongly related to land use, with improved pasture and intensive grazing occurring on class 4 land, moderate grazing pressure on class 5 land, and low grazing intensities on class 6 land.

Opportunity cost was calculated after determination of the optimum configuration of paddocks in the study area to achieve particular conservation goals, at both the property and regional scales.[2]

Opportunity cost differed little between four different analyses, with the property scale outcomes costing marginally more than the regional scale outcomes. The addition of threatened species to the criteria had little impact (see Table 7.1). When the calculation of the total opportunity cost was done on the basis of including all land in the

Table 7.1. Results of optimisation process for different criteria for nature conservation at two scales

Criteria	Vegetation targets only		Vegetation and threatened species targets	
Scale	Property	Regional	Property	Regional
Opportunity cost of selected units	42 501	40 849	42 749	41 674
Opportunity cost of native vegetation contributing to targets in selected units	25 254	24 388	25 406	25 043
Native vegetation area in selected units (ha)	38 468	37 329	38 488	37 407
Unselected native vegetation on land of no opportunity cost (ha)	8334	9043	7301	8968
Total native vegetation selected + unselected native vegetation on land of no opportunity cost (ha)	46 802	46 372	46 789	46 375
% of study area that should remain as native vegetation if criteria applied	74.27	73.58	74.24	73.59
Capture of threatened species	187	198	251	255

units, rather than just the areas within the units needed to attain conservation targets, there was a 50 per cent increase in the score in all four analyses (as shown in Table 7.1).

All sets of criteria resulted in a large proportion of threatened species targets being captured (see Table 7.1). However, a small proportion of the targets for threatened species were not met by selections based only on vegetation criteria, as reflected in the slightly larger opportunity costs required to cater for threatened species targets on top of the vegetation targets (see Table 7.1).

Fenced management units selected to contribute to one target also appear in the other sets. Two hundred and thirty-nine were selected in at least one set, with 177 of these selected in all four analyses.

The addition of all of the land in capability class 6 that was not captured by the selections would maintain almost three-quarters of the study area under native vegetation (Table 7.1). However, in all four analyses approximately 5500 ha of class 4 and 5 land under native vegetation was included in the selection of units, indicating a potential for substantial impacts on the conservation values of the region under a business as usual scenario. Two of the nine private properties could not satisfy the criterion of 30 per cent native vegetation, while the study area as a whole could.

Selection for vegetation communities resulted in good representation of threatened species, largely because reserves, including informal reserves on private land, were automatically included in the final set, and these have high concentrations of threatened species by design. In addition, threatened species are often concentrated in vegetation communities that have been severely depleted in area, and are of high conservation significance, such as *Themeda triandra* grassland. Although sampling for threatened species in the study area has not been systematic or comprehensive, it would be foolish not to ensure that known occurrences were in the selected set.

If opportunity cost mapping is valid, so that there is no economic reason to intensify land use in areas with class 6 land capability, the maintenance of present management in the fenced management units that we selected would result in most of the study area remaining under native vegetation (Table 7.1). The potential conflict between nature conservation and economic development narrows down to approximately 5500 ha out of 63 000 ha.

The total opportunity cost of the threatened species and vegetation selection at either the property or landscape scale is unlikely to be huge in dollar terms, given that to increase the intensity of land use in areas currently occupied by native vegetation would involve substantial initial investment in clearing and continuing investment in fertiliser application (see Table 7.2). For example, if the opportunity cost of the area of native vegetation necessary to meet the targets within units selected at a property scale, using all criteria, was paid at June 2004 prices and costs, the annual cost would be approximately AUD$635 000. An economic analysis that properly discounted both returns and improvement costs to give a net present opportunity cost would therefore be unlikely to result in a politically unattainable figure. However, the figure could become substantially larger if alternative, more intensive, land uses than sheep grazing became an option for class 4 or 5 land.

As restoration is prohibitively expensive, and often difficult, in environments altered for agriculture, and as property boundaries are an ecologically arbitrary unit of analysis, it seems more appropriate to plan for conservation at the landscape scale than the property scale. However, property scale planning is also needed. Certification of property plans is mooted to become the basis of satisfaction of legislative requirements under planning and nature conservation acts in some Australian jurisdictions, including Tasmania. Landowners have an obvious interest in minimising the opportunity costs incurred by their satisfaction of any legislative requirements. If governments are to provide any payment for the service of managing land in a way that could sacrifice potential profits, they also have an obvious interest in achieving their nature conservation goals at least cost.

The type of analysis documented above may be useful, at the property and regional scales, in many parts of the world. The GIS data necessary to undertake the analyses

Table 7.2. Calculating the opportunity cost of maintaining land use in native vegetation

Land capability class	**4**	**5**	**6**
Predominant grazing intensity	High	Medium	Low
Potential return if developed to full potential (D)	10	5	1
Potential return if undeveloped, allowing low impact grazing (U)	3	2	1
Opportunity cost index (D – U)	7	3	0

for the Upper Macquarie catchment are available in most parts of Australia, and many other parts of the world. The technology needed to undertake the analyses is simple, inexpensive and readily available.

Using nature in marketing

If wool prices continue to decline, the runs may be soon depopulated of sheep, except for those subsidised to provide nature conservation services. Two million sheep have already gone from Tasmania (see Chapter 1). Technological fixes, such as self shearing sheep, may reduce the cost of producing wool, although the scope for substituting capital for labour is very low, with Australian Bureau of Statistics data for 2004 recording only the equivalent of three or four people, including the grazier, their spouse and contractors, employed on a typical wool-growing property. A more promising option may be to sell the clean natural environment, not the bale.

In the richer countries of the world there is a strong market for products that are bereft of contamination by artificial chemical compounds and that are composed of raw materials that have been produced in an environmentally responsible manner. A romantic story-line also helps. The New Zealanders sell woollen garments using a debatable environmental harmony story-line and the spectacular mountain scenery of some of their sheep grazing country. As one Tasmanian grazier said:

> 'In New Zealand they are small, compact, but they have a culture of being a highly successful industry because they have nothing else. New Zealand excels at what they do because they are forced to. Australia is diverse. Tasmania is similar to New Zealand.'

In the United States there are chains of large stores that only sell certified organic products. Producers of woollen garments can gain a major market and price advantage if they can attach even the most prosaic of products, like socks, to a credible clean and green story-line. Indeed, a sock manufacturer from Colorado visited Tasmania recently to assess the potential for Tasmanian wool to contribute to such a story-line, and subsequently used Tasmanian wool. A major Tasmanian wool-broking firm is working to develop environmental markets.

Some Tasmanian graziers do not like the idea of wool being sold as anything else than wool:

> 'Big risks differentiating one from another – should be **selling wool**.'

> 'Farmers are competitive against each other. We can't afford to do this in terms of wool. Need to market the product.'

> 'Got to get people using wool. Got to be careful saying this is natural wool. Wool is natural, full stop.'

Others have reservations about the practicalities of selling the fizz rather than the drink:

> 'I don't think there's potential there. I think it (environmental accreditation) is a lovely idea but I don't think it's going to put one more cent on the wool. Will it sell more wool? In a business, you have to sell a product. You can have the best wool in the world, but I don't think it will bring one more cent in. I would happily do it though.'

> '(Environmental) accreditation is all a bit touchy feely. If there are no extra dollars in the bank, then we're not into it.'

> 'More work to get there and not getting anything extra – now the volume is down and there's not as much product to offer.'

> 'Accredition, yes, if they pay and create a premium. It probably will happen but not soon.'

> 'Market is a steep hill. TQW (Tasmanian Quality Wool), plenty on board for a long time but no-one recognising it and giving them a fair run.'

> 'For marketing, volume is the problem. Need to be able to supply market.'

> 'There's a danger that overseas customers will demand products that they see as sustainable agriculture. I don't agree that in business the customer is always right. The customer is an ignorant fool.'

> 'Problem is what consumers think they'd like rather than about farmers doing the right thing. Establishing criteria will be quite difficult.'

Yet others see environmental accreditation as desirable:

> 'Want to lead in criteria, improve landscapes, improve way we grow wool for a better product, access markets.'

> 'Part of trying to grow a successful wool-growing business. Decrease costs, increase farm management, increase markets.'

> 'Believe you can do. See opportunities and take them. It's up to you and you have to be prepared to change. Traditional methods handicap people – need to look at new concepts.'

Some forms of environmental accreditation of wool are beginning to have a history. TQW is a quality control system, valued by some graziers:

> 'TQW, quality fibre, what you see is what you get, chemical-free or recording of chemicals used is optional, tick a box. Main contamination is coloured fibre, most costly, next most costly is foreign fibre (e.g. baling twine).'

> 'TQW, edge with wool if under that scheme. Whether you get a premium or not is another thing. TQW accreditation hasn't meant that we get a better price.'

Others are more dubious about it:

> 'I'm not in TQW until I build a new shed.'

> 'Some of the compliance requirements are ridiculously stupid.'

The quality control in the TQW system is for the product, not the environmental probity of the process of producing it at all stages. The organic scheme, Demeter, is more comprehensive:

> 'CSIRO do the chemical testing on the wool – they keep excellent records of all wool provided. Wool is tested on farm then on tops (processing plant) – most contaminants are removed with wool grease.'

One grazier, who had been supplying wool for this label since 1994 said: 'I should be more protective of my own market.' Demeter wool qualifies under the European Union (EU) Ecolabel, along with wool subject to less stringent conditions. The grazier describes the Demeter process and rewards:

> 'Wool supplier has to sign declaration no chemicals used on wool, all way down processing chain. Declare what you use and what you haven't. Are getting premium in small niche market. Set price for last seven years has been stable. Growers get premium of 20 per cent (range 10–40 per cent). The premium is obtained through a unique product delivered to market as they want it.'

Some graziers perceive the requirements of some buyers to be too severe:

> 'Europeans won't buy unless some of these things (accreditation conditions) are in place. There's a problem with super, you wouldn't run half the sheep without fertiliser. I do lime the country, lime helps to make P more available. It needs to be there.'

> 'Talking about chemicals, there's a contract out now from a European company buying wool. They want 20 micron wool. Roberts rang, but said "We've got to ask you a few questions", and one question was "Do you super the country". "Yes" and they wouldn't buy the wool that came off.'

The judgement that 'the customer is an ignorant fool' is one that relates to the difficulties in satisfying all of what is perceived by many graziers to be a 'politically correct' agenda. Not the least problem in this regard is coping with demands for improved animal welfare from organisations such as People for the Ethical Treatment of Animals (PETA), particularly the demand for the abolition of mulesing, an operation that removes fleece and skin from around the anus of the sheep. Mulesing is done to prevent fly strike, a more protracted unpleasant experience for a sheep than the operation itself, but not necessarily one to which many sheep would be subjected. Some graziers would prefer strong action to repel PETA:

> 'The government should be standing up to it.'
>
> 'Can't afford to give in to extremist groups, once you do where do you stop?'

Others accept that there is not much to be done about this sort of market pressure, except comply:

> 'Issues fly strike, mulesing. Can be problem if not on to it. Need to manage for it.'
>
> 'Mulesing – can get away without it, but better chance of not getting fly strike (with mulesing) – won't be the end of the world if we can't mules.'

Elimination of surgical mulesing by 2010 is now the aim of the industry.

Tasmanian wool growers are in a better position to take marketing advantage of the 'politically correct' agenda than most other Australian and New Zealand wool producers. It is less difficult in Tasmania than most other places to adopt a chemical-free approach to wool production, at least from the runs. The Tasmanian run country provides an excellent environmental story. Threatened animal and plant species, and other natives with much customer appeal, are prominent in the run country ecosystems, which appear to have as close to a physically sustainable agricultural system as traditional paddy rice agriculture. Many native animals that are common on the Tasmanian runs have been rendered extinct on the mainland. New Zealand pastures are riddled with exotics, have erosion problems and retain few of the native birds and reptiles (there were no native terrestrial mammals in New Zealand when the Maori arrived 700 years ago). The Tasmanian Government promotes a 'clean, green, high quality' image for the produce of the state, an image partly backed by reality. If accreditation for environment-friendly and biodiversity-friendly wool is going to work anywhere to help save an industry, it is going to work for wool growing in Tasmania.

Between 2002 and 2005 a group of wool producers, scientists, wool brokers and nature conservation bureaucrats[3] sat around various tables in the historic towns and properties of the Midlands of Tasmania trying to make progress towards developing criteria and processes that could be used for the biodiversity aspects of environmental accreditation. The eventual consensus on the criteria was that they should be outcome-oriented and simply measured: no net decrease in native vegetation cover and quality; and maintenance or increase of populations of threatened species dependent on a property. The eventual consensus on process was that accreditation should be based on the biodiversity component of property plans. Such a plan was prepared for one property to test the process, which seemed feasible.

A felicitous convergence?

At the time of writing (May 2006), controls on clearance of non-forest vegetation in Tasmania seem most likely to be through the mechanism of approved property plans. There is the potential for plans produced under such a process to also serve the function of environmental accreditation. As one grazier said in relation to rare and threatened species (RATS): 'RATS can become an opportunity, rather than threats.' There is also increasing interest at all levels of government in the concept of stewardship payments for environmental services, although the initial trial projects funded by the Australian Government do not conform to the suggestions made above, in that they do not target areas of high conservation significance.

The outcomes of discussions of the Midlands group have also formed input to several other processes aimed at improving the environmental and economic performance of the wool industry, including a statewide attempt to incorporate environmental management systems into farming systems (FARMSAT), the adoption of a property planning approach to sustainable farm management (NRM North, NRM Cradle Coast), and supporting wool producers who wish to target a niche market for ecologically sustainable wool production (Roberts Ltd). The approach of a biodiversity management plan was also well received from international wool buyers (Teko Socks, USA) and the Southern Australian Beef Research Council. As part of a deliberative social conversation, it is hoped they will contribute to solving the many problems of the wool producers and the nature on which they partially rest.

Notes

1 This particular grazier also said, in mitigation: 'will fence off five times as much'.

2 Digital data were used for mapped vegetation communities, threatened species locations, aerial photos, cadastral information and modelled land capability. These data sets, provided by the Department of Primary Industries, Water and Environment, Tasmania, are used in state planning processes. Property boundaries in the catchment were obtained from cadastral maps and created as new layers in a GIS database. Within these properties, fenced management units that could be identified from aerial photographs were mapped. Further information on management units and their boundaries, in addition to information on grazing regimes for each unit, was gained during interviews with land owners/managers. Six hundred and twenty-four units were mapped. The fenced management unit was used because this is the scale at which management is implemented on the farms, so these boundaries potentially correspond to some environmental patterns (such as vegetation cover, land capability), and any future management decisions are most easily implemented at this level. The area occupied by each vegetation community and land capability class in each fenced management unit was calculated. Threatened species were added as attributes to the units in which they were found. A threatened species score was derived for each fenced management unit. Every endangered species in a unit was given a score of three, each listed endemic plant species a score of three, each non-endemic vulnerable plant a score of two, and each non-endemic plant classified as rare and at risk a score of one. The area of each modelled land capability class in each fenced management unit was calculated. Modelled land capability was used to calculate the potential opportunity cost of conservation for each fenced management unit. On the basis of the literature (e.g. Garden *et al.* 2000; Leonard and Kirkpatrick 2004; Kirkpatrick, Gilfedder *et al.* 2005; Chapter 5) it was assumed that the native vegetation of the Northern Midlands can support low intensity and/or carefully managed land uses, such as sheep grazing at low continuous stocking rates, or with spelling during flowering and seeding periods. This assumption formed the basis for calculation of the opportunity cost of grazing land (see Table 7.2). The potential return figures (D) for each class are approximate and relative, based on stocking

rates able to be supported by the different land classes if developed to their full potential. These figures were determined from the results of a study by Friend *et al.* (2001) and from interviews with landholders (the stocking rates supported on different areas of their properties). For example, the bush runs (predominantly class 6) present on a number of the properties in the study area are largely set stocked, at approximately one dry sheep equivalent per hectare per year (dse/ha/year). If superphosphate is applied to native pasture on class 5 land, carrying capacity increases from 2–3 dse/ha/year to approximately 4–6 dse/ha/year (Friend *et al.* 2001). The more fertile soils that have been converted to improved pasture (largely class 4) support approximately 8–12 dse/ha/year. Class 6 land is not appropriate for development and can only support 1 dse/ha/year. There is therefore no opportunity cost in maintaining present land use on class 6 land. There are substantial differences between potential carrying capacity after development and carrying capacity on unmodified native pasture for class 4 and 5 land (see Table 7.2). The opportunity cost index (D – U) is the difference in dse/ha/year between developing land to its full potential and maintaining it as native vegetation. In June 2004, the gross margin for a dse/ha/year was approximately AUD$25.00 for wool production properties. This figure varies markedly through time, responding to supply and demand conditions. A relative cost for each fenced management unit in the study area was calculated by multiplying the opportunity cost index by the area occupied by each land class. For example, fenced management unit 'X' is 20 ha, which consists of 15 ha of class 5 land, and 5 ha of class 6 land, so the relative opportunity cost of fenced management unit X is (15*3)+(5*0) = 45. If a fenced management unit contained land devoted to improved pasture or crops, this area was given an opportunity cost index of zero in these calculations. Two thresholds for conservation criteria were applied to the study area: firstly, to maintain 100% of the area of vegetation communities of very high significance, 0% of communities adequately reserved at the statewide level, and 15% of remaining vegetation communities; secondly, these targets for vegetation communities, plus all locations of endangered, or vulnerable Tasmanian endemic, plants, and 50% of the locations of non-endemic rare and at risk and vulnerable plants. Some papers and reviews have nominated a threshold of 30% native vegetation cover of a system for, among other things, the maintenance of animal populations (e.g. Andrén 1994), and general maintenance of ecological processes (e.g. McIntyre *et al.* 2002), so this was applied as an additional target in all analyses. As we could not obtain statewide pre-European extent of all woodland and grassland communities mapped in the study area, the analysis was done based on the current extent of vegetation communities in the study area. Of course this is not optimal, considering that most of these communities have been severely depleted since European arrival, and that further clearing will reduce targets. Effective conservation requires the persistence of all species, so analyses were done including the known locations of threatened species in the study area. These thresholds for conservation criteria were applied at the individual property scale, and at the regional scale. Optimisation was used to select fenced management units to meet targets for conservation, whilst keeping the opportunity cost as low as possible. The numbers of fenced management units were sufficiently low that approximate approaches to optimisation, such as simulated annealing (e.g. Andelman *et al.* 1999) were not needed. True optimisation could be used in all analyses as there was no need to include those fenced management units that did not contain target species or more than 1 ha of target communities, or those units that had to be selected because they contained species or communities for which total occurrence was needed to reach conservation targets, or those that already had some form of protection as natural vegetation. If any targets for any particular analysis were not reached by units in the last two classes, the combination of remaining fenced management units with the lowest opportunity cost that attained or exceeded all these targets was added to the set. This selection was done from calculations of the total opportunity cost and degree of target attainment for each possible combination of these remaining fenced management units, easily accomplished in a spreadsheet. If a threatened plant occurred in a fenced management unit that was greater than 20 ha in area, and that was not required to contribute to other vegetation targets, we placed a buffer of 252 metres around the plant, and only included this buffer in the set. This figure was used by us because it creates a circle of 20 ha in area, a critical threshold for bird species richness identified by MacDonald and Kirkpatrick (2003). The data used in the analyses have some limitations. The occurrences of threatened species were not mapped in a systematic way; therefore there is bias towards easily accessible sites and certain species. Some fenced management unit boundaries may have been mapped at too coarse a scale, and some may be inaccurate or missing. The land capability layer is modelled, and not all areas have been groundtruthed. As the layer was created at a scale of 1:100 000, the minimum area that can be adequately depicted on the map represents approximately 64 hectares on the ground (Noble 1992). Further details of these analyses can be found in Jensen (2004).

3 Those who participated in more than one meeting of the group were: Rae Glazik (DPIWE), Lindsay Young (wool producer), Simon Foster (wool producer), Henry Foster (wool producer), Julian von Bibra (wool producer), Julian Cotton (wool producer), Rod Headlam (wool producer), Eric Hutchinson (Roberts Wool), Leanne Sherriff (University of Tasmania), Jamie Kirkpatrick (University of Tasmania), Kerry Bridle (University of Tasmania), Meg Nichols (wool producer), Shauna Ellis (wool producer), John Atkinson (wool producer), David Atkinson (wool producer), Tom Dunbabin (wool producer), Andrew Johnston (DPIWE), Sarah Ackland (wool producer), Steve Barrington (wool producer) and Andrew Hamlet (DPIWE).

Chapter 8

Some overall conclusions

J.B. Kirkpatrick and K.L. Bridle

Summary

Several lessons are drawn from the book as a whole: the virtues of diversity; the risks attached to innovation; the virtues of cooperation; and, the potential for all people, sheep and nature to prosper on the runs.

Conclusions

This book was not designed as a moral tale, or a political program, although it may seem to the reader to have ended up that way. Its authors commenced their research with an attachment to nature, a curiosity about natural history and the fate of nature on the runs, and a desire to help develop ways for nature to coexist with wool producers and their sheep. The authors ended up supplementing their sympathy for nature with empathy for both wool producers and sheep, both of which seem to have much harder lives than researchers, the lives of whom are hard enough.

A lot of fascinating particularities have been reported. This idiographic approach, as it is called by philosophers of science, enables a rich perception of the reality it addresses, but does not lead us far on the path to cosmic theories, or even universal generalisations. In parts of several of the chapters the authors have taken a different approach. In normative investigations, they attempt to put ideas to the sword by testing hypotheses, albeit little ones. This final section is the place for the big picture lessons that have been drawn from the story of people, sheep and nature in Tasmania.

The high value of diversity is a prime lesson that can be drawn from the story. At the economic level, diverse solutions for making money from sheep on diverse blocks of land have been documented. The value of risk spreading within an enterprise is also apparent. There seems little doubt that inter-property diversity in management approaches leads to better overall economic outcomes than any one type of solution, and that intra-property diversity in income streams mitigates the effects of fluctuations in prices of particular commodities and services. At a social level, wool producers differ in their opinions on appropriate management procedures and goals, their preparedness to innovate and take risks, and their attachments to various aspects of their properties, including various elements of nature. This social diversity creates a complex spatial matrix of different patterns of management of the runs, which, in turn, results in a habitat complexity that allows native species with different requirements to survive, and sometimes prosper, at the landscape scale. If the runs were all managed in the one way, native biodiversity would suffer. The high species diversity of the run ecosystems makes them more resilient to environmental perturbation than human-created exotic ecosystems, contributing to their economic value as a drought reserve and for shelter. Species that might not necessarily be regarded positively by graziers may be missed when they are gone, as in the case of that corpse-disposer, the Tasmanian devil.

A second lesson that can be drawn from the story is the economic, social and environmental risk attached to widely adopted innovations, especially in the absence of prior assessment of potential consequences. The use of gorse, South African boxthorn and hawthorn for hedging seemed like a good idea at the time, as did the introduction of the rabbit. The introduction of Vermont genes into the Tasmanian Merino also turned out not to be a good idea. In more recent times, tree clearance, combined with intensive irrigation of crops using pivot irrigators, may eventually reap a salty harvest. We do not possess the scientific knowledge to know whether this will be a serious problem, or not. The virtue of voluntary and enforced cooperation can be argued to be a third lesson, exemplified by the successes and failures in tackling the diseases and pests of the runs.

The final lesson is that it is possible to maintain or improve nature conservation values in runs used by sheep, sometimes in surprising ways, as with the importance of heavy grazing in the maintenance of populations of some threatened plant species, and with the potential for combining stewardship payments with fuzzy regulation. The futures of people, sheep and nature on the runs of Tasmania are uncertain, but the potential exists to benefit them all.

Appendixes

Appendix 1

The percentage frequency of comments related to grazing management

Regime	
Rotational grazing	48
Set-stocked	35
Rest runs in spring to allow growth/seeding	31
Combination of set-stocked and rotational grazing	25
Young wethers not on runs/on best runs	15
Rest runs in summer to allow growth	15
Urea/lick/supplements used in dry time or to encourage browsing of less palatable species	15
Lick/urea blocks no good/don't use	15
Use native pasture all year	12
Use native pasture in winter	12
Try to spell some runs for reserve	12
Wethers in run all year	10
Use urea/lick blocks for runs	10
Stocking rate varies with different types of country	10
Rest runs for variable periods	10
Rest native pasture in winter	8
Young and old wethers on better runs	6
Use improved after shearing	6
Use cell grazing	6
Lamb onto improved	6
Feed ewes prior to lambing	6
Use native pasture in summer	4
Cattle and sheep run together	4
Variable stocking rate on different runs	4
Use improved to build up body fat/fatten wethers for sale	4
Urea/lick supplements used in winter	4
Manage according to soil types	4

Regime (continued)	
Graze improved in summer	4
Young ewes on native (cells)	2
Use rough country before shearing	2
Use native pasture in autumn	2
Use native pasture as feedlot during drought	2
Use bush as flood feed reserve	2
Rest runs for long period after drought	2
Rest runs after shearing	2
Rest runs by changing stocking rate	2
Rest runs in lush years	2
High altitude country used as extended spring	2
Graze bush intensively for short period	2
Don't use bush for stock	2
Don't put cattle in runs in droughts	2
Bush as summer feed reserve	2
Stock heavily in early summer for short period	2
Wethers onto stubble in summertime	2
Wethers on improved most of year	2
Don't run hoggets on improved	2
Destock in dry years	2
Stock type	
Saxon Merinos	17
Breed of sheep on bush run important	6
Corriedale x Merino	6
Border Cross	6
Merinos	4
White Suffolk	2
Polwarth Cross	2
Cormo	2
Purposes	
Sagg/tussock country shelter for lambing	25
Use bush for shelter	23
Use bush for feed	23
Use all/part of native pasture as drought reserve	23
Sagg/tussock country shelters off-shears	19
Use bush as drought reserve	17
Cattle used to knock back tussock	12
Native pasture as shelter	12
Focus on superfine from native pasture	12

Purposes (continued)	
Better grass cover in runs keeps wool cleaner	12
Avoid dusty conditions/maintain ground cover	8
Use bush for off-shears	6
Need improved for nutrition/live weight gain	6
Use fodder (hay, grain etc.) to let pasture get away after bad season	4
Rest runs in spring to improve stock condition	4
Cattle used to open up bush	4
Cattle not used to knock back tussocks	4
Stock shelter in bush in bad weather	2
Runs not used for shelter	2
Rotating maintains biomass/reduces erosion	2
Rotation in bush prevents it from getting overgrown	2
Cell grazing allows eucalypt regeneration	2
Want to maintain stock through winter	2
Urea blocks aid mustering	2
Stock heavily to keep micron low	2
Lower stocking rates improve health	2
Easterly aspect used as shelter for lambing	2
Biodynamic management has reduced health risks	2
Other uses on property	
Cropping	33
Irrigating	23
Growing focus on prime lamb	12
Cattle in native/bush prior to calving	4
Cattle on improved	2
Variety of crops grown	2
Growing focus on horticulture/cropping	2
Fodder grown	2
Recognition of useful native plants	
Wallaby grass	25
Kangaroo grass	25
Silver tussock	17
Tussock	10
Poa	10
Themeda	8
Weeping grass	4
Danthonias	4
Blue grass tussock	4
Spear grass	2

Recognition of useful native plants (continued)	
Schoenus	2
Poa rodwayi	2
Native herbs	2
Microlaena	2
Observations and opinions	
Runs cleaner/fewer parasite problems/fewer inputs	25
Increasing/want to increase stocking rate on native country	12
Fine wool from runs more consistent (fewer breaks) staple	12
Wool from runs almost/completely chemical-free	12
Sheep prefer improved	8
Cattle better suited to native pasture than sheep	4
Native grasses valued for production, not biodiversity	4
Biggest cost of bush is mustering (time)	4
Strategy = pasture focused	4
Cell grazing has a risk of footrot	4
Runs improve resilience of farm to rainfall variability	2
People overvalue native pasture	2
Not enough information on good management of native grasses	2
Sheep in poor condition get worse on runs	2
Muster problems on bush runs	2
Maintaining cover important for health	2
Cattle not suited to run country	2
Bush has low/zero input costs	2
Stock losses are factor in using bush	2
Sheep prefer to be in runs	2
Sheep moving from improved to native will eat saplings	2
Kangaroo grass is an indicator of sward health	2
Abundance of kangaroo grass indicates insufficient grazing pressure	2
Kangaroo grass important	2
Prefer open grazing to woodland	2
Other	
Recent changes to management	15
Little change in management over the years	10
Have decreased stocking rate on native	8
Increasing focus of graziers on irrigation effects stock management	4
Reducing stock numbers didn't reduce cash flow	4
Moving away from wethers (fine wool)	4
Keep grazing charts	4
Lease/agist into extra land at times of year	4

Other (continued)	
Usually agist	2
Water an issue in runs	2
Sheep behaviour indicator of rotation period	2
Bush wethers wear coats to keep clip clean	2
Sheep not in bush when flowers there	2
Spring flush in improved causes micron blow-out	2
Shearing timed from autumn break	2
Sheep grouped on basis of feed requirements and mating groups	2
Shear when expecting break in wool	2
Periodic problems with salinity of stock water	2
Older sheep present fibre risk	2
Feed sheep as part of normal management	2
Feed sheep during dry periods	2
Don't keep grazing charts	2
Decrease in clip micron during drought helps balance decrease in weight	2

Appendix 2

Percentage frequency of non-production-related comments on native species/communities

Values	
Like aesthetics of trees/bush	44
Value birdlife	19
Value bush as habitat	6
Value observation of ecosystem processes	4
Value biodiversity	4
Recreational value	4
Protecting bush makes farm sustainable	4
Bush has conservation value	4
Wilderness value	2
Want to leave place in better state	2
Interstate and international visitors appreciate bush	2
Like native pasture aesthetics	2
Value frogs	2
Bush is good for kids growing up	2
Bush as skyline reserve	2
Bush as 'well-being, security'	2
Trees	
Dieback an issue	35
Thickening of woody vegetation a problem	15
Dieback caused by combination of factors	12
Dieback not an issue	6
Concern about tree dieback	6
White gums more susceptible to dieback	6
Dieback caused/contributed to by possums	6
Aerial spray has killed trees	6
Major changes in vegetation over last century	4
Dieback problem getting better	4
Try not to damage trees with burns	2
Dieback not caused by stock	2
Insufficient funding and research on dieback	2
Concern about tree death from fires	2
Slow tree growth in last 30 years	2
Hakea a nursery plant for *Eucalyptus gunnii*	2
Green blackwoods (in bush)	2
Fire might reduce insect attack on trees	2
Cool fires prevent eucalypt regeneration	2
Other native plants	
Wallaby grass present	25

Other native plants (continued)	
Kangaroo grass present	25
Silver tussock present	17
Tussock present	10
Poa present	10
Themeda present	8
Wattles present	6
Weeping grass present	4
Danthonia present	4
Blue grass tussock present	4
Spear grass present	2
Schoenus present	2
Rare plants in riparian zone	2
Poa rodwayi present	2
Orchids present	2
Native legume present	2
Native herbs present	2
Microlaena present	2
Lost much tussock grasses over recent decades	2
Lilysag present	2
Bush has changed little over the years	2
Coastal wattle present	2
Native animals	
Decrease in devil population	25
Increase in possums	12
Cats a problem	10
Wombats present	10
Increase in native hens	10
Quolls present	8
Increase in cats	8
Increase in wedgetails	8
Increase in wombats	8
Decrease in bandicoots	8
Devils present	6
Decrease in waterbird population	6
Brown tussock moth present	6
Bandicoots present	6
Concerned about birds nesting with burning	4
Potoroos present	4
Increase in wood ducks	4
Increase in wallabies	4
Devils gone	4

Native animals (continued)	
Decrease in possums	4
Bettongs present	4
Platypus present	2
Increase in rosellas	2
Increase in roos	2
Increase in hawks	2
Increase in black ducks	2
Increase in devil population	2
Galaxids present	2
Forester kangaroos present	2
Decrease in native hens	2
Decrease in snake numbers	2
Decrease in roo population after drought	2
Decrease in quolls	2
Decrease in ground nesting birds	2
Decrease in wattle birds	2
Decrease in green parrots	2
Cats have destroyed birdlife	2
Actions	
Regenerating parts of bush	35
Fenced off some bush/remnants	29
Covenant on area/some of area	17
Extensive tree planting	8
Manage parts of bush for conservation	8
Regional Forest Agreement reserve present	6
Fenced off some hilltop	6
Plans to fence off some/more bush	4
Planting shelter belts/want to plant shelter belts	4
Private Forests Reserve Program reserve present	4
Hot fires threaten biodiversity	2
Don't use bush for stock	2
Sheep not in bush when flowers there	2
Covenant on native grassy vegetation	2
Area/s of native grassland/grassy woodland fenced off permanently	2
Planting trees in low altitude salt-prone areas	2

Glossary

Many of the definitions follow those in the glossary of Mokany *et al.* (2006).

Accelerated soil erosion – soil erosion to a greater degree than would have occurred without human-caused disturbance.

Acre – an archaic measure of area. One acre equals 0.405 hectares.

Annual species – a species whose life cycle is completed within one year (typically: germination in autumn, flowering and seed set in early spring, senescence and death of the plant in mid–late spring). Examples include annual grasses (e.g. *Vulpia*), annual clovers (e.g. subterranean clover), and broadleaf species (e.g. capeweed).

Block grazing – see 'Rotational grazing' and 'Cell grazing'.

Carrying capacity – the maximum stocking rate that can be supported by a pasture without causing a deterioration in the condition of that pasture. The carrying capacity of a pasture is not static, but varies between seasons and years.

Cell grazing – stock are moved around a 'cell' consisting of a large number of paddocks of a size which permits high stocking densities to be applied to each paddock for a suitably short period. The formal application of this grazing strategy requires pasture growth rates to be monitored, and grazing periods to be calculated according to growth and recovery rates of the desirable pasture species.

Continuous grazing – see 'Set stocking'.

Dryland salinity – the result of a process that brings salt to the surface on naturally dry land.

Dry sheep equivalents (DSE) – a standard unit used for estimating the feed requirements of different classes of livestock. The basic unit (1 DSE) is the Merino wether (35–45 kg liveweight) maintained in store condition by an intake of feed providing 7 MJ/day of metabolisable energy.

Ewe – female sheep in reproductive age group.

Exotic species – (also called non-native or introduced species) those species that were not present in Tasmania before European settlement.

Forage value – relates to the overall quality of a pasture as a source of nutrients for stock (i.e. digestibility, protein content, metabolisable energy).

Forbs – herbaceous plants other than grasses, sedges and rushes.

Grassland – vegetation where the groundcover is dominated by grasses, and tree cover is sparse or non-existent.

Grazing intensity – the amount of forage that is eaten in a set time in a set area.

Grazing pressure – the relationship between the number of grazing animals and the amount of forage that is available at a particular time (DSE per kg of pasture dry matter).

Growing points – areas of the plant that actively produce new cells and plant tissue.

Herb species – herbaceous plant species that do not have woody stems. Includes grasses, broadleaf species, sedges, rushes and lilies.

Hogget – female sheep too young to have reproduced.

Introduced species – see 'Exotic species'.

Land capability – the suitability of land for economic use in a sustainable fashion.

Mob stocking – where a pasture is grazed by a large number of stock (high stocking rate) for usually a short period of time (also called 'crash grazing').

Native species – those species that were present in Tasmania before European settlement.

Non-native species – see 'Exotic species'.

Opportunity cost – the rewards forgone by not developing less the cost of developing.

Overgrazing – when stocking rates exceed the carrying capacity of a pasture. Overgrazing results in degradation in the condition of a pasture, including loss of desirable pasture species, weed invasion and soil erosion.

Palatability – relates to the preferences stock show for consuming one plant (or plant part) over another. Palatability can be affected by plant characteristics (such as protein content), as well as animal requirements and preferences.

Pasture – refers to an area that is devoted to the production of forage for stock, where the ground cover is dominated by herbaceous species (e.g. grasses, broadleaf species).

Perennial species – plant species that persist for more than two growing seasons. Examples include most native grasses (e.g. wallaby grass, kangaroo grass), many native broadleaf species (e.g. common everlasting), and most sown grasses (e.g. perennial ryegrass, cocksfoot).

Ram – uncastrated male sheep of reproductive age.

Resting – sheep are taken off the run.

Rotational grazing – grazing management involving regular intervals of grazing and spelling of a paddock.

Runs – the parts of a property that are largely covered by native species and are used to graze or shelter stock.

Sagg – *Lomandra longifolia*.

Scats – defecatory pellets, as with sheep dung or wallaby droppings.

Sedges – 'grass-like' plants. However, they are not true grasses (Poaceae), but rather in the family Cyperaceae.

Set stocking – grazing management where stock remain in a paddock for long periods of time. The paddock may be spelled for part of the year, and the stocking rate may change somewhat over time.

Spelling – where all the stock are removed from a paddock for a period of time.

Stocking rate – the number of stock per unit area, often expressed as dry sheep equivalents (DSE) per hectare.

Time control grazing – see 'Rotational grazing'.

Transhumance – the seasonal movement of stock between areas of different climate.

Umbrageous – umbrella-shaped.

Utilisation – the proportion of the forage produced by a pasture that is eaten by stock.

Volunteer species – an exotic plant species that has not been deliberately sown or introduced, but has dispersed and spread itself to different areas. Examples include annual grass weeds, broadleaf weeds and small leaf annual clovers.

Wether – castrated male sheep.

References

Abensberg-Traun M, Smith GT, Arnold GW and Steven DE (1996). The effects of habitat fragmentation and livestock grazing on animal communities in remnants of gimlet *Eucalyptus salubris* woodland in the Western Australian wheatbelt. I. Arthropods. *Journal of Applied Ecology* **33**, 1281–1301.

Allan K (1998). A dendroclimatological investigation of *Phyllocladus aspleniifolius* (Labill.) Hook. f. PhD thesis, University of Tasmania.

Andelman S, Ball I, Davis F and Stoms D (1999). *SITES V 1.0: An Analytical Toolbox for Designing Ecoregional Conservation Portfolios.* The Nature Conservancy, Boise, Idaho.

Andrén H (1994). Effects of habitat fragmentation on birds and mammals with different proportions of suitable habitat: a review. *Oikos* **71**, 355–366.

Archer S (1995). Tree grass dynamics in a prosopis–thornscrub savanna parkland – reconstructing the past and predicting the future. *Ecoscience* **2**, 83–99.

Australian Bureau of Agricultural and Resource Economics (2006). *Financial Performance of Wool Producing Farms to 2004–05.* Australian Wool 06/1.

Banks MR (1973). *The Lake Country of Tasmania: A Symposium Conducted by the Royal Society of Tasmania at Poatina, Tasmania, November 11–12, 1972.* Royal Society of Tasmania, Hobart.

Bartley TB (1855). *Golden Fields for Tasmania: A Few Words as to the Value and Importance of Peruvian Guano, as a Manure: Addressed to the Farmers of the Colony.* Charles Wilson, Launceston.

Bell EA (1965). *Roberts, Stewart and Co Ltd: An Historic Centenary.* Mercury Press, Hobart.

Bennett AF and Ford LA (1997). Land use, habitat change and the conservation of birds in fragmented rural environments: a landscape perspective from the Northern Plains, Victoria, Australia. *Pacific Conservation Biology* **3**, 244–261.

Bennett AF, Lumsden LF and Nicholls AO (1994). Tree hollows as a resource for wildlife in remnant woodlands: spatial and temporal patterns across the northern plains of Victoria, Australia. *Pacific Conservation Biology* **1**, 222–235.

Bennett LT (1994). The expansion of *Leptospermum laevigatum* on the Yanakie Isthmus, Wilson's Promontory, under changes in the burning and grazing regimes. *Australian Journal of Botany* **42**, 555–564.

Binning CE (1997). Beyond reserves: options for achieving nature conservation objectives in rural landscapes. In *Frontiers in Ecology – Building the Links.* (Eds N Klomp and I Lunt) pp. 155–168. Elsevier, Oxford.

Bischoff J (1832). *Sketch of the History of Van Diemen's Land, Illustrated by a Map of the Island, and an Account of the Van Diemen's Land Company.* John Richardson, London.

Bob Reid and Associates (2003). 'Performance indicators Tasmania results 2000–01 and 2001–02.' Report to 8 x 5 Wool Profit Program, Hobart.

Bond WJ (2005). Large parts of the world are brown or black: a different view on the 'green world' hypothesis. *Journal of Vegetation Science* **16**, 261–266.

Bond WJ and Keeley JE (2005). Fire as a global 'herbivore': the ecology and evolution of flammable ecosystems. *Trends in Ecology and Evolution* **20**, 387–393.

Bowman DMJS, Maclean AR and Crowden RK (1986). Vegetation–soil relations in the lowlands of south-west Tasmania. *Australian Journal of Ecology* **11**, 141–153.

Bowman DMJS, Walsh A and Milne DJ (2001). Forest expansion and grassland contraction within a *Eucalyptus* savanna matrix between 1941 and 1994 at Litchfield National Park in the Australian monsoon tropics. *Global Ecology and Biogeography Letters* **10**, 535–548.

Boyce J (2006). Canine revolution: the social and environmental impact of the introduction of the dog to Tasmania. *Environmental History* **11**, 102–129.

Bridle KL and Kirkpatrick JB (1997). Local environmental correlates of variability in the organic soils of moorland and alpine vegetation, Mt Sprent, Tasmania. *Australian Journal of Ecology* **22**, 196–205.

Bridle KL and Kirkpatrick JB (1998). Why do tall herbs rarely dominate Tasmanian alpine vegetation? Evidence from islands in the Ouse River system. *Papers and Proceedings of the Royal Society of Tasmania* **132**, 9–14.

Bridle KL and Kirkpatrick JB (1999). The comparative effects of stock and wild vertebrate grazing on treeless subalpine vegetation, eastern Central Plateau, Tasmania. *Australian Journal of Botany* **47**, 817–834.

Bridle KL and Kirkpatrick JB (2001). Impacts of grazing by vertebrate herbivores on the flower stem production of tall alpine herbs, eastern Central Plateau, Tasmania. *Australian Journal of Botany* **49**, 459–470.

Bridle KL, Kirkpatrick JB, Cullen P and Shepherd RR (2001). Recovery in alpine heath and grassland following burning and grazing, eastern Central Plateau, Tasmania. *Arctic, Antarctic and Alpine Research* **33**, 348–356.

Briggs JM, Knapp AA and Brock BL (2002). Expansion of woody plants in tallgrass prairie: a fifteen year study of fire and fire-grazing interactions. *American Midland Naturalist* **147**, 287–294.

Brown JR and Archer S (1999). Shrub invasion of grassland: recruitment is continuous and not regulated by herbaceous biomass or density. *Ecology* **80**, 2385–2396.

Brown JR and Carter J (1998). Spatial and temporal patterns of exotic shrub invasion in an Australian tropical grassland. *Landscape Ecology* **13**, 93–102.
Brown MJ and Podger FD (1982). On the apparent anomaly between observed and predicted percentages of vegetation types in south-west Tasmania. *Australian Journal of Ecology* **7**, 3–12.
Busby JR (1988). Potential implications of climatic change on Australia's flora and fauna. In *Greenhouse: Planning for Climate Change.* (Ed GI Pearman) pp. 387–398. CSIRO, Melbourne.

Calder J (1840). 'Lots at Central Plains and Nineteen Lagoons'. Nomenclature Board of Tasmania, Reference D7.
Cale P and Hobbs RJ (1991). Condition of roadside vegetation in relation to the nutrient status. In *Nature Conservation 2: The Role of Corridors.* (Eds DA Saunders and RJ Hobbs) pp. 353–362. Surrey Beatty and Sons, Sydney.
Campbell A (1994). *Landcare: Communities Shaping the Land and the Future.* Allen and Unwin, Sydney.
Campbell Town District High School (2002). *Living on the Land: Glimpses of a Community Working with Wool.* Campbell Town District High School, Campbell Town.
Chappell K (2000). Aquatic macrophytes of the Macquarie and South Esk rivers, Tasmania. MEnvMgt thesis, University of Tasmania, Hobart.
Clarke MF and Schedvin N (1999). Removal of bell miners *Manorina melanophrys* from *Eucalyptus radiata* forest and its effect on avian diversity, psyllids and tree health. *Biological Conservation* **88**, 111–120.
Clements A (1983). Suburban development and resultant changes in the vegetation of the bushland of the northern Sydney region. *Australian Journal of Ecology* **8**, 307–319.
Close DC and Davidson NJ (2004). Review of rural tree decline in a changing Australian climate. *Tasforests* **15,** 1–18.
Cluzeau D, Binet F, Vertes F, Simon JC, Riviere JM and Trehen P (1992). Effects of intensive cattle trampling on a soil-plant-earthworms system on two grassland types. *Soil Biology and Biochemistry* **24**, 1661–1665.
Coblentz BE (1978). The effects of feral goats (*Cabra hircus*) on island ecosystems. *Biological Conservation* **13**, 279–286.
Collins SL, Knapp AK, Briggs JM, Blair JM and Steinauer EM (1998). Modulation of diversity by grazing and mowing in native tallgrass prairie. *Science* **280**, 745–747.
Colwell JD (1965). An automatic procedure for the determination of phosphorus in sodium carbonate extracts of soils. *Chemistry and Industry* May 1965, 893–895.
Commonwealth of Australia (2001). *Australian Native Vegetation Assessment. National Land and Water Resources Audit.* Commonwealth of Australia, Canberra.
Cooke BD (1987). The effects of rabbit grazing on regeneration of she-oaks, *Allocasuarina verticillata*, and saltwater ti-trees, *Melaleuca halmaturorum*, in the Coorong National Park, South Australia. *Australian Journal of Ecology* **13**, 11–20.

Costella DA, Lunt ID and Williams JE (2000). Effects of invasion by the indigenous shrub *Acacia sophorae* on plant composition of coastal grasslands in south-eastern Australia. *Biological Conservation* **96,** 113–121.
Cottle DJ (1991). *Australian Sheep and Wool Handbook*. Inkata Press, Melbourne.
Crawford G (1966). The Scotts: Thomas, George and James. *Papers and Proceedings of the Tasmanian Historical Research Association* **14**, 12.
Cremer KW (1990). *Trees for Rural Australia*. Inkata Press, Melbourne.
Cubit S (1996). Burning back with the snow: traditional approaches to grassland management in Tasmania. *Australian Geographical Studies* **43**, 216–224.
Cubit S (1998). The Walls of Jerusalem: a grazing history and an alpine heritage. *Papers and Proceedings of the Tasmanian Historical Research Association* **45**, 203–214.
Curr E (1824). *An Account of the Colony of Van Diemen's Land, Principally Designed for the Use of Emigrants*. London.
Curtis A (1997). Landcare, stewardship and biodiversity conservation. In *Frontiers in Ecology: Building the Links* (Eds N Klomp and I Lunt) pp. 143–154. Elsevier, Oxford.

Daley EA and Kirkpatrick JB (2005). Native riparian vegetation in Tasmania. *Cunninghamia* **8**, 409–430.
Daniels GD and Kirkpatrick JB (2006). Does variation in garden characteristics influence the conservation of birds in suburbia? *Biological Conservation* **133**, 326–335.
Davidson J (1938). On the growth of the sheep population in Tasmania. *Transactions of the Royal Society of South Australia* **62**, 342–346.
Day G and Jessup J (1984). *The History of the Australian Merino*. William Heineman, Victoria.
Dickinson KJM and Kirkpatrick JB (1986). The impact of grazing pressure in clearfelled, burned and undisturbed eucalypt forest. *Vegetatio* **66**, 133–136.
Diez S and Foreman P (1997). 'The management of grasslands on the riverine plain'. Department of Natural Resources and Environment, Melbourne.
Dodson JR and Mooney SD (2002). An assessment of historic human impact on south-eastern Australian environmental systems, using late Holocene rates of environmental change. *Australian Journal of Botany* **50**, 455–464.
Dorrough J, Ash J and McIntyre S (2004). Plant responses to livestock grazing frequency in an Australian temperate grassland. *Ecography* **27**, 798–810.
Dorrough J, Yen A, Turner V, Clark SG, Crosthwaite J and Hirth JR (2004). Livestock management and biodiversity conservation in Australian temperate grassy landscapes. *Australian Journal of Agricultural Research* **55**, 279–295.
Dow DD (1977). Indiscriminate interspecific aggression leading to almost sole occupancy of space by a single species of bird. *Emu* **77**, 115–121.
Duncan F (1990). Eucalypts in Tasmania's changing landscape. *Tasforests* **2**, 202–208.

Eckhardt CH, Wilgen WB, Biggs CH (2000). Trends in woody vegetation cover in the Kruger National Park, South Africa, between 1940 and 1998. *African Journal of Ecology* **38**, 108–115.

Edwards I (1973). Management of water yield. In *The Lake Country of Tasmania: A Symposium Conducted by the Royal Society of Tasmania at Poatina, Tasmania, 11–12 November 1972.* (Ed MR Banks) pp. 177–182. Royal Society of Tasmania, Hobart.

Edwards J (1998). Changes in attitudes and intentions towards remnant bush in subhumid Tasmania 1992–1998. BSc (Hons) thesis, School of Geography and Environmental Studies, University of Tasmania, Australia.

Elkington TT (1981). Effects of excluding grazing animals from grassland on sugar limestone in Teesdale, England. *Biological Conservation* **20**, 25–35.

Ellis RC and Thomas I (1998). Pre-settlement and post-settlement vegetational change and probable Aboriginal influences in a highland forested area in Tasmania. In *Australia's Ever Changing Forests: Proceedings of the First National Conference on Australian Forest History, Canberra, 9–11 March 1998.* (Eds KJ Frawley and N Semple) pp. 199–216. Department of Geography and Oceanography, Australian Defence Force Academy, Campbell, ACT.

Evandale Historical Society and the Clarendon Committee (1992). *Clarendon and its People: A Brief History.* Evandale Historical Society, Evandale.

Fensham RJ (1985). The pre-European vegetation of the Midlands in Tasmania. BSc (Hons) thesis, Geography Department, University of Tasmania, Hobart.

Fensham RJ (1989). The pre-European vegetation of the Midlands of Tasmania: a floristic and historical analysis of vegetation patterns. *Journal of Biogeography* **16**, 29–45.

Fensham RJ (1991). Management plan case study: Sherwood Bush, Epping Forest, Tasmania. In *Tasmanian Native Bush: A Management Handbook.* (Ed JB Kirkpatrick) pp. 162–183. Tasmanian Environment Centre, Hobart.

Fensham RJ and Fairfax RJ (2002). Aerial photography for assessing vegetation change: a review of applications and the relevance of findings for Australian vegetation history. *Australian Journal of Botany* **50**, 415–429.

Fensham RJ, Holman JE and Cox MJ (1999). Plant species responses along a grazing disturbance gradient in an Australian grassland. *Journal of Vegetation Science* **10**, 77–86.

Fensham RJ and Kirkpatrick JB (1989). The conservation of original vegetation remnants in the Midlands, Tasmania. *Papers and Proceedings of the Royal Society of Tasmania* **123**, 229–246.

Fensham RJ and Kirkpatrick JB (1992). The eucalypt forest-grassland/grassy woodland boundary in central Tasmania. *Australian Journal of Botany* **40**, 123–138.

Fitzgerald M (2004). Responses of macroinvertebrate communities to altered management regimes in grassy woodland and heathy open-forest, Tasmania. BSc

(Hons) thesis, School of Geography and Environmental Studies, University of Tasmania.

Foreman P (1999). Grazing as a conservation management tool at Terrick Terrick National Park: a new grassland reserve in the Victoria Riverina. In *Proceedings of a Conference on Managing Grassy Ecosystems.* (Eds V Craigie and C Hocking) pp. 75–88. Victoria University, St Albans.

Francis R (1997). Effects of sheep grazing on the invertebrate communities of Liaweenee Moor. BSc (Hons) thesis, School of Geography and Environmental Studies, University of Tasmania.

Freemark KE, Boutin C and Keddy CJ (2002). Importance of farmland habitats for conservation of plant species. *Conservation Biology* **16**, 399–412.

Freudenberger D and Ozolins A (2000). Isolated trees under threat from farming. *Farming Ahead* **104**, 38–39.

Friend DA, Cameron AS, Povey AJ and Dolan PL (1997). Seed banks in a natural pasture in Tasmania, Australia: implications for species composition change. In *Proceedings of the 18th International Grasslands Congress.* Section 28, pp. 9–10.

Friend DA, Dolan PL and Hurst AM (1999). Effect of spelling on the growth of a native pasture in Tasmania, Australia. In *Proceedings of the VI International Rangeland Congress, Volume 1.* (Eds D Eldridge and D Freudenberger) pp. 489–491. Sixth International Rangelands Conference Inc.

Friend DA, Dolan PL and Hurst AM (2001). Superphosphate boosts productivity and profitability of native grass pastures in Tasmania. In *Proceedings of the 10th Australian Agronomy Conference, Hobart, Tasmania, Australia.*

Friend DA, Thompson RP, Ball PD, Corkrey SR and Graham BRF (1997). 'The pastures of the Midlands and Derwent Valley, Tasmania: relationships to environment and management factors'. Final report to the International Wool Secretariat, Project DAT 44. Tasmanian Institute of Agricultural Research, Launceston.

Garden DL, Lodge GM, Friend DA, Dowling PM and Orchard BA (2000). Effects of grazing management on botanical composition of native grass-based pastures in temperate south-east Australia. *Australian Journal of Experimental Agriculture* **40**, 225–245.

Gibbons P and Boak M (2002). The value of trees for regional conservation in an agricultural landscape. *Ecological Management and Restoration* **3**, 205.

Gibson CWD, Hambler C and Brown VK (1992). Changes in spider (Araneae) assemblages in relation to succession and grazing management. *Journal of Applied Ecology* **29**, 132–142.

Gibson N and Kirkpatrick JB (1985). Vegetation associated with localised snow accumulation at Mount Field West, Tasmania. *Australian Journal of Ecology* **10**, 91–99.

Gibson N and Kirkpatrick JB (1989). Effects of the cessation of grazing on the grasslands and grassy woodlands of the Central Plateau, Tasmania. *Australian Journal of Botany* **37**, 55–63.

Gibson W (1887). *The Tasmanian Merinos of Scone*. W. Gibson & Son, Scone.

Gilfedder L and Kirkpatrick JB (1993a). 'Factors influencing the condition of natural vegetation remnants in subhumid Tasmania'. A report to the Australian Nature Conservation Agency, University of Tasmania, Hobart.

Gilfedder L and Kirkpatrick JB (1993b). Germinable soil seed and competitive relationships between a rare native species and exotics in a semi-natural pasture in the Midlands, Tasmania. *Biological Conservation* **64**, 113–119.

Gilfedder L and Kirkpatrick JB (1994a). Climate, grazing, disturbance, and the population dynamics of *Leucochrysum albicans* at Ross, Tasmania. *Australian Journal of Botany* **42**, 417–430.

Gilfedder L and Kirkpatrick JB (1994b). Culturally induced rarity? – the past and present distributions of *Leucochrysum albicans* in Tasmania. *Australian Journal of Botany* **42**, 404–416.

Gilfedder L and Kirkpatrick JB (1995a). 'A survey of landholder attitudes and intentions towards the long-term conservation of native lowland grasslands'. A report from the University of Tasmania to the Grasslands Ecology Unit, Australian Nature Conservation Agency.

Gilfedder L and Kirkpatrick JB (1995b). Improving the conservation of vegetation remnants in Tasmania. In *Nature Conservation 4: The Role of Networks* (Eds DA Saunders, JL Craig and EM Mattiske) pp. 613–621. Surrey Beatty and Sons, Sydney.

Gilfedder L and Kirkpatrick JB (1996). The distribution, ecology and conservation needs of *Colobanthus curtisiae* West. *Papers and Proceedings of the Royal Society of Tasmania* **130**, 25–30.

Gilfedder L and Kirkpatrick JB (1997). The influence of owners' intentions on remnant bush conservation in Tasmania. In *Conservation Outside Nature Reserves* (Eds P Hale and D Lamb) pp. 217–225. Centre for Conservation Biology, The University of Queensland, Brisbane.

Gilfedder L and Kirkpatrick JB (1998a). Factors influencing the integrity of remnants in subhumid Tasmania. *Biological Conservation* **84**, 89–96.

Gilfedder L and Kirkpatrick JB (1998b). Distribution, disturbance tolerance and conservation of *Stackhousia gunnii* Hook. f. in Tasmania. *Australian Journal of Botany* **46**, 1–13.

Gilfedder L, Kirkpatrick JB, Wapstra A and Wapstra H (2003). *The Nature of the Midlands*. Midlands Bushweb, Longford.

Gott B (2005). Aboriginal fire management in south-eastern Australia: aims and frequency. *Journal of Biogeography* **32**, 1203–1208.

Gouldthorpe J and Gilfedder L (2002). 'Biodiversity strategy case study: Tasmanian Northern Midlands bioregion, a report to the national Land and Water Resources Audit'. Nature Conservation Branch Technical Report 02/08, Department of Primary Industries, Water and Environment, Tasmania.

Greenwood K and McKenzie B (2001). Grazing effects on soil physical properties and the consequence for pastures: a review. *Australian Journal of Experimental Agriculture* **41**, 1231–1250.

Grey MJ, Clarke MF and Loyn RH (1997). Initial changes in the avian communities of remnant eucalypt woodlands following a reduction in the abundance of noisy miners, *Manorina melanocephala*. *Wildlife Research* **24**, 631–648.

Grey MJ, Clarke MF and Loyn RH (1998). Influence of the noisy miner, *Manorina melanocephala*, on avian diversity and abundance in remnant grey box woodland. *Pacific Conservation Biology* **4**, 55–69.

Grice MS (1995). 'An assessment of soil and land degradation in Tasmania'. Hobart. Department of Primary Industries and Fisheries, Tasmania and the National Soil Conservation Program.

Grose CJ and Moreton RM (1996). 'Land capability survey of Tasmania. South Esk report'. Department of Primary Industries, Water and Environment, Tasmania.

Groves RH, Kelaitis K and Moore CWE (1973). Relative growth of *Themeda australis* and *Poa labillardierei* in pots in response to phosphorus and nitrogen. *Australian Journal of Botany* **21**, 1–11.

Harris S and Kitchener A (2005). *From Forest to Fjaeldmark: Descriptions of Tasmania's Vegetation*. Department of Primary Industries, Water and Environment, Hobart.

Harwood CE and Jackson WD (1975). Atmospheric losses of four plant nutrients during a forest fire. *Australian Journal of Forestry* **38**, 92–99.

Heinjus D (1992). *Farm Tree Planting*. Inkata Press, Adelaide.

Henderson W and Wilkins CW (1975). The interaction of bushfires and vegetation. *Search* **16**, 130–133.

Hennessy JT, Gibbens RP, Tromble JM and Cardenas M (1983). Vegetation changes from 1935 to 1980 in Mesquite dunelands and former grasslands of southern New Mexico. *Journal of Range Management* **36**, 370–374.

Hester A and Hobbs RJ (1991). Influence of fire and soil nutrients on native and non native annuals at remnant vegetation edges in the Western Australian wheatbelt. *Journal of Vegetation Science* **3**, 101–108.

Historical Committee of the National Trust of Australia (Tasmania) (1966). *Campbell Town, Tasmania*. Campbell Municipal Council, Campbell Town.

Hitchins FE (1950). *Tangled Skeins – A Historical Survey of Australian Wool Marketing*. Robertson and Mullens, Melbourne.

Hobbs RJ and Huenneke LF (1992). Disturbance, diversity and invasion: implications for conservation. *Conservation Biology* **6**, 324–337.

Hogg A and Kirkpatrick JB (1974). The phytosociology and synecology of some southern Tasmanian eucalypt forests and woodlands. *Journal of Biogeography* **1**, 227–245.

Howe HF and Westley LC (1988). *Ecological Relationships of Plants and Animals.* Oxford University Press, New York.

Hutchinson KJ and King KL (1970). Sheep numbers and soil arthropods. *Search* **1**, 41–42.

Hutchinson LM, Swain R and Driessen M (2001). *Snakes and Lizards of Tasmania.* Fauna of Tasmania Handbook No. 9. University of Tasmania and Department of Primary Industries, Water and Environment, Hobart.

Jackson ML (1958). *Soil Chemical Analysis.* Prentice Hall, New Jersey.

Jackson WD (1968). Fire, air, water and earth – an elemental ecology of Tasmania. *Proceedings of the Ecological Society of Australia* **3**, 9–16.

Jackson WD (1973). Vegetation. In *The Lake Country of Tasmania: A Symposium Conducted by the Royal Society of Tasmania at Poatina, Tasmania, 11–12 November 1972.* (Ed MR Banks) pp. 61–85. Royal Society of Tasmania, Hobart.

Jensen A (2004). Selecting least-cost areas for conservation in rural landscapes. BSc (Hons) thesis, School of Geography and Environmental Studies, University of Tasmania.

Jetson T (1989). *The Roof of Tasmania – A History of the Central Plateau.* Pelion Press, Launceston.

Kadmon R and Harari-Kremer R (1999). Studying long-term vegetation dynamics using digital processing of historical aerial photographs – Israel and Jordan. *Remote Sensing of Environment* **68**, 164–176.

Kee S (1990). 'Midlands archaeological Aboriginal site survey'. Occasional Paper No. 26. Department of Parks, Wildlife and Heritage, Tasmania.

Kellaway RG (1989). Geographical change in Tasmanian agriculture during the Great Depression. PhD thesis, University of Tasmania, Hobart.

King KL and Hutchinson KJ (1983). The effects of sheep grazing on invertebrate numbers and biomass in unfertilized natural pastures of the New England Tableland (NSW). *Australian Journal of Ecology* **8**, 245–255.

King KL, Hutchinson KJ and Greenslade P (1976). The effects of sheep numbers on Collembola in sown pastures. *Journal of Applied Ecology* **13**, 731–739.

Kintsch JA and Urban DL (2002). Focal species, community representation, and physical proxies as conservation strategies: a case study in the Amphibolite Mountains, North Carolina, USA. *Conservation Biology* **16**, 936–947.

Kirkpatrick JB (1975). Phytosociological analysis of the vegetation of Lagoon Beach, Tasmania. *Papers and Proceedings of the Royal Society of Tasmania* **109**, 53–63.

Kirkpatrick JB (1983). An iterative method for establishing priorities for the selection of nature reserves: an example from Tasmania. *Biological Conservation* **25**, 354–372.

Kirkpatrick JB (1986a). The viability of bush in cities – ten years of change in an urban grassy woodland. *Australian Journal of Botany* **35**, 691–708.

Kirkpatrick JB (1986b). Tasmanian alpine biogeography and ecology and interpretation of the past. In *Flora and Fauna of Alpine Australasia: Ages and Origins.* (Ed BA Barlow) pp. 229–242. CSIRO Publishing, Collingwood.

Kirkpatrick JB (1991). The magnitude and significance of land clearance in Tasmania in the 1980s. *Tasforests* **3**, 11–14.

Kirkpatrick JB (1998). Nature conservation and the regional forest agreement process. *Australian Journal of Environmental Management* **5**, 31–37.

Kirkpatrick JB (1999). *A Continent Transformed: Human Impact on the Vegetation of Australia*. 2nd ed. Oxford University Press, Sydney.

Kirkpatrick JB (2003). Natural history. In *The Nature of the Midlands.* (Eds L Gilfedder, JB Kirkpatrick, A Wapstra and H Wapstra). pp. 43–57. Midlands Bushweb, Longford.

Kirkpatrick JB (2004). Vegetation change in an urban grassy woodland 1974–2000. *Australian Journal of Botany* **52**, 597–608.

Kirkpatrick JB, Barker P, Brown MJ, Harris S and Mackie R (1995). 'The Reservation Status of Tasmanian Vascular Plant Communities'. Wildlife Scientific Report 95/4, Parks and Wildlife Service, Tasmania.

Kirkpatrick JB, Bridle KL, Leonard SWJ and Gilfedder L (2005). Managing sheep for nature conservation on wool-growing properties – some preliminary observations on the short-term impacts of grazing regime changes. In *Grassland Conservation and Production – Both Sides of the Fence.* (Eds C O'Dwyer and S Hamilton) pp. 100–105. *Proceedings of the 4th Stipa Conference on Management of Native Grasses and Pastures 11–13 October 2005, Burra, SA*, FLFR University of Melbourne, Dookie Campus.

Kirkpatrick JB and Brown MJ (1984). The palaeogeographic significance of local endemism in Tasmanian higher plants. *Search* **15**, 112–113.

Kirkpatrick JB and Dickinson KJM (1982). Recent destruction of natural vegetation in Tasmania. *Search* **13**, 186–187.

Kirkpatrick JB and Dickinson KJM (1984). *Vegetation of Tasmania, 1:500,000.* Forestry Commission, Tasmania, Hobart.

Kirkpatrick JB and Gibson N (1999). Towards an explanation of the altitudinal distributions of three species of *Eucalyptus* in central Tasmania. *Australian Journal of Ecology* **24**, 123–141.

Kirkpatrick JB and Gilfedder L (1991). Characteristics and management requirements of threatened and unreserved native plant species on private land in Tasmania. In *Tasmanian Native Bush – A Management Handbook.* (Ed JB Kirkpatrick) pp. 213–269. Tasmanian Environment Centre, Hobart, Tasmania.

Kirkpatrick JB and Gilfedder L (1995). Maintaining integrity compared with maintaining rare and threatened taxa in remnant bushland in subhumid Tasmania. *Biological Conservation* **74**, 1–8.

Kirkpatrick JB and Gilfedder L (1998). Conserving weedy natives: two Tasmanian endangered herbs in the Brassicaceae. *Australian Journal of Ecology* **23**, 466–473.

Kirkpatrick JB and Gilfedder L (1999). *Tasmanian Bushcare Toolkit*. Department of Primary Industries, Water and Environment, Hobart.

Kirkpatrick JB, Gilfedder L, Bridle KL and Zacharek A (2005). The positive and negative conservation impacts of sheep grazing and other disturbances on the vascular plant species and vegetation of lowland subhumid Tasmania. *Ecological Management and Restoration* **6**, 51–60.

Kirkpatrick JB, Gilfedder L and Fensham R (1988). *City Parks and Cemeteries: Tasmania's Grassy Heritage*. Tasmanian Conservation Trust, Hobart.

Kirkpatrick JB and Harris S (1995). 'The conservation of Tasmanian dry coastal vascular plant communities'. Wildlife Scientific Report 95/1, Parks and Wildlife Service, Tasmania.

Kirkpatrick JB and Harwood CE (1983a). The conservation of Tasmanian wetland macrophytic species and communities. *Papers and Proceedings of the Royal Society of Tasmania* **117**, 5–20.

Kirkpatrick JB and Harwood CE (1983b). Tasmanian wetland plant communities. *Australian Journal of Botany* **31**, 437–451.

Kirkpatrick JB and Marks F (1985). Observation of drought damage to some native plant species in eucalypt forests and woodlands near Hobart, Tasmania. *Papers and Proceedings of the Royal Society of Tasmania* **119**, 15–21.

Kirkpatrick JB and Nunez M (1980). Vegetation-radiation relationships in mountainous terrain: eucalypt-dominated vegetation in the Risdon Hills. *Journal of Biogeography* **7**, 197–208.

Kirkpatrick JB, Zacharek A and Chappell K (2000). Testing methods for mitigation of tree dieback in Tasmanian dry eucalypt forests and woodlands. *Pacific Conservation Biology* **6**, 94–101.

Kleijn D and Sutherland WJ (2003). How effective are European agri-environment schemes in conserving and promoting biodiversity? *Journal of Applied Ecology* **40**, 947–969.

Knapp AA (1996). Cheatgrass (*Bromus tectorum* L.) dominance in the Great Basin Desert: history, persistence and influences of human activities. *Global Environmental Change* **6**, 37–52.

Knapp AA, Blair JM, Briggs JM, Collins SL, Hartnett DC, Johnson LC and Towne EG (1999). The keystone role of bison in American tallgrass prairie. *Bioscience* **49**, 39–50.

Knapp AA and Soule PT (1998). Recent *Juniperus occidentalis* (Western Juniper) expansion on a protected site in western Oregon. *Global Change Biology* **4**, 347–357.

Kutiel P, Cohen O, Shoshany M and Shub M (2003). Vegetation establishment on the southern Israeli coastal sand dunes between the years 1965 and 1999. *Landscape and Urban Planning* **67**, 141–156.

Landsberg J and Wylie FR (1983). Water stress, leaf nutrients and defoliation: a model of dieback of rural eucalypts. *Australian Journal of Ecology* **8**, 27–41.

Lange RT and Graham CR (1983). Rabbits and the failure of regeneration in Australian arid zone. *Australian Journal of Ecology* **8**, 377–381.

Leahy PJ, Tibby J, Kershaw AP, Heijnis H and Kershaw JS (2005). The impact of European settlement on Bolin Billabong, a Yarra River floodplain lake, Melbourne, Australia. *River Research and Applications* **21**, 131–149.

Leigh JH, Wimbush DJ, Wood DH, Holgate MD, Slee AV, Stanger MG and Forrester RI (1987). The effects of rabbit grazing and fire on a subalpine environment. I. Herbaceous and shrubby vegetation. *Australian Journal of Botany* **35**, 433–464.

Leith P (2006). *Conversations about Climate: Seasonal Variability and Grazier's Decisions in the Eastern Rangelands.* University of Tasmania, Hobart.

Leonard SWJ and Kirkpatrick JB (2004). Effects of grazing management and environmental factors on native grassland and grassy woodland, Northern Midlands, Tasmania. *Australian Journal of Botany* **52**, 529–542.

Lester S (1994). *Spring Bay, Tasmania – A Social History.* Artemis Publishing and Marketing Consultants, Hobart.

Lindenmayer D, Claridge A, Hazell D, Michael D, Crane M, MacGregor C and Cunningham R (2003). *Wildlife on Farms: How to Conserve Native Animals.* CSIRO Publishing, Collingwood.

Lobo JM, Hortal J and Cabrero-Sanudo FJ (2006). Regional and local influence of grazing activity on the diversity of a semi-arid dung beetle community. *Diversity and Distributions* **12**, 111–123.

Lockwood M, Walpole S and Miles CA (2000). 'Economics of remnant native vegetation conservation on private property'. National Research and Development Program on Rehabilitation, Management and Conservation of Remnant Vegetation, Research Report 2/00, Land and Water Resources Research and Development Corporation, Canberra.

Loofs-Samorzewski M (2003). The impacts of edge effects and other disturbances on Tasmanian coastal heathlands. PhD thesis, School of Geography and Environmental Studies, University of Tasmania, Hobart.

Lowndes AG and Maze WH (1937). *Land Utilization Regions of Tasmania.* University of Sydney Publications in Geography No 4. University of Sydney, New South Wales.

Lunt ID (1990). Impact of an autumn fire on a long-grazed *Themeda triandra* (kangaroo grass) grassland: implications for management of invaded, remnant vegetation. *Victorian Naturalist* **107**, 45–51.

Lunt I (1997). Effects of long-term vegetation management on remnant grassy forests and anthropogenic native grasslands in south-eastern Australia. *Biological Conservation* **81**, 287–297.

Lunt ID and Morgan JW (1999). Vegetation changes after 10 years of grazing exclusion and intermittent burning in a *Themeda triandra* (Poaceae) grassland reserve in south-eastern Australia. *Australian Journal of Botany* **47**, 537–552.

MacArthur RH and Wilson EO (1967). *The Theory of Island Biogeography.* Princeton University Press, Princeton.

MacDonald M (2001). Habitat fragmentation and its effects on birds and grasshoppers in eucalypt remnants in the Tasmanian Midlands. PhD thesis, School of Geography and Environmental Studies, University of Tasmania, Hobart.

MacDonald M and Kirkpatrick JB (2003). Explaining bird species composition and richness in eucalypt-dominated remnants in subhumid Tasmania. *Journal of Biogeography* **30**, 1415–1426.

McIntyre S and Barrett GW (1992). Habitat variegation, an alternative to fragmentation. *Conservation Biology* **6**, 146–147.

McIntyre SJ, McIvor G and Heard KM (2002). *Managing and Conserving Grassy Woodlands.* CSIRO Publishing, Collingwood.

McIntyre S, McIvor JG and MacLeod ND (2000). Principles for sustainable grazing in eucalypt woodlands: landscape-scale indicators and the search for thresholds. In *Management for sustainable ecosystems.* (Eds P Hale, A Petrie, D Moloney and P Sattler) pp. 92–100. Centre for Conservation Biology, University of Queensland, Brisbane.

McIntyre S and Tongway D (2005). Grassland structure in native pastures: links to soil surface condition. *Ecological Management and Restoration* **6**, 43–50.

McKay A (1962). *Journals of the Land Commissioners for Van Diemen's Land 1826–28.* University of Tasmania in conjunction with the Tasmanian Historical Research Association, Hobart.

McNaughton SJ (1976). Serengeti migratory wildebeest: facilitation of energy flow by grazing. *Science* **191**, 92–94.

McNaughton SJ (1984). Grazing lawns: animals in herds, plant form, and coevolution. *American Naturalist* **124**, 863–866.

McNaughton SJ and Tarrants JL (1983). Grass leaf silification: natural selection for an inducible defence against herbivores. *Proceedings of the National Academy of Sciences, USA* **80**, 790–791.

McQuillan PB (1999). The effects of changes in Tasmanian grasslands on the geometrid moth tribe Xanthorhoini (Geometridae: Larantiinae). In *The Conservation and Biodiversity of Invertebrates.* (Eds W Ponder and D Lunney) pp. 121–128. Royal Zoological Society of New South Wales, Mosman.

Madany MH and West NE (1983). Livestock grazing–fire regime interactions within montane forests of Zion National Park. *Ecology* **64**, 661–667.

Margules CR and Pressey RL (2000). Systematic conservation planning. *Nature* **405**, 243–253.

Massy C (1990). *The Australian Merino.* Penguin Books, Ringwood.

Mendel LC (1999). Scenery to wilderness: national park development in Tasmania. PhD thesis, University of Tasmania, Hobart.

Mendel LC (2002). The consequences for wilderness conservation in the development of the national park system in Tasmania, Australia. *Australian Geographical Studies* **40**, 71–83.

Mendel LC and Kirkpatrick JB (1999). Assessing temporal changes in the reservation of the natural aesthetic resource using pictorial content analysis and a grid-based scoring system – the example of Tasmania. *Landscape and Urban Planning* **43**, 181–190.

Mendel LC and Kirkpatrick JB (2002). Historical progress of biodiversity conservation in the protected-area system of Tasmania, Australia. *Conservation Biology* **16**, 1–11.

Meredith LA (1852). *My Home in Tasmania*. John Murray, London.

Mills AJ and Fey MV (2005). Interactive response of herbivores, soils and vegetation to annual burning in a South African savanna. *Austral Ecology* **30**, 435–444.

Minchin PR (2001). *DECODA: Database for Ecological Community Data Version 3.00.* ANU, Canberra.

Minitab Inc. (1998). *User's Guide 2: Data Analysis and Quality Tools Release 12.* Minitab Inc, USA.

Mitchell A (1962). 'Report of the soil conservation problems on the Central Plateau and South Esk River catchment in Tasmania'. Report to the Standing Committee on Conservation of Natural Resources of Tasmania.

Mokany K, Friend D, Kirkpatrick JB and Gilfedder L (2006). *Managing Tasmanian Native Pastures – A Technical Guide for Graziers.* TIAR, Hobart.

Moore RM (1970). *Australian Grasslands.* Australian National University Press, Canberra.

Moretto AS and Distel RA (1999). Effects of selective defoliation on the competitive interaction between palatable and unpalatable grasses native to a temperate semi-arid grassland of Argentina. *Journal of Arid Environments* **42**, 167–175.

Morgan S (1992). *Land Settlement in Early Tasmania – Creating an Antipodean England.* Cambridge University Press, Melbourne.

Muller S (2002). Diversity of management practices required to ensure conservation of rare and locally threatened plant species in grasslands: a case study at a regional scale (Lorraine, France). *Biodiversity and Conservation* **11,** 1173–1184.

Mysterud A, Hansen LO, Peters C and Austrheim G (2005). The short-term effect of sheep grazing on selected invertebrates (Diptera and Hemiptera) relative to other environmental factors in an alpine ecosystem. *Journal of the Zoological Society of London* **266**, 411–418.

National Land and Water Resources Audit (2001). 'Australian Native Vegetation Assessment 2001'. NLWRA, Turner, ACT.

Neyland M (1996). *Tree Decline in Tasmania.* Land and Water Management Council, Hobart.

Noble KE (1992). *Land Capability Survey of Tasmania – Land Capability Handbook*. Department of Primary Industry, Tasmania.

Nunez M (1983). Estimation of solar radiation received on slopes in Tasmania. *Papers and Proceedings of the Royal Society of Tasmania* **117**, 153–159.

Oates N and Clarke B (1987). *Trees for the Back Paddock*. Goddard and Dobson, Box Hill, Victoria.

Oliver I, Beattie AJ and York A (1998). Spatial fidelity of plant, vertebrate and invertebrate assemblages in multiple-use forest in eastern Australia. *Conservation Biology* **12**, 822–835.

Panzer R and Schwartz M (1998). Effectiveness of a vegetation-based approach to insect conservation. *Conservation Biology* **12**, 693–702.

Parsonson IM (1998). *The Australian Ark: A History of Domesticated Animals in Australia*. CSIRO Publishing, Collingwood.

Pharo EJ, Beattie AJ and Pressey RL (2000). Effectiveness of using vascular plants to select reserves for bryophytes and lichens. *Biological Conservation* **96**, 371–378.

Pharo EJ, Kirkpatrick JB, Gilfedder L, Mendel L and Turner PAM (2005). Predicting bryophyte diversity in grassland and eucalypt-dominated remnants in subhumid Tasmania. *Journal of Biogeography* **32**, 2015–2024.

Phillips YA (1987). *Bartley of 'Kerry Lodge': A Portrait of a Pioneer in Van Diemen's Land*. Blackwall, Tasmania.

Pink K (2003). *Winds of Change – A History of Woolnorth*. The Van Diemen's Land Company, Timaru, NZ.

Pinkard L (1992). Ecological impediments to the establishment of trees by direct seeding on pastoral sites in the Midlands of Tasmania. MEnvSt thesis, University of Tasmania.

Plomley NJB (1966). *Friendly Mission*. THRA, Hobart.

Potts BM, Potts WC and Kantvilas G (2001). The Miena cider gum, *Eucalyptus gunnii* ssp. *divaricata*: a taxon in rapid decline. *Papers and Proceedings of the Royal Society of Tasmania* **135**, 57–62.

Prevost J (1988). *From the Epping Banks to the Esk*. Foot and Playsted, Launceston.

Prober SM and Thiele KR (1995). Conservation of the grassy white box woodlands: relative contributions of size and disturbance to floristic composition and diversity of remnants. *Australian Journal of Botany* **43**, 349–366.

Prober SM, Thiele K and Lunt I (2002). Identifying ecological barriers to restoration in temperate grassy woodlands: soil changes associated with different degradation states. *Australian Journal of Botany* **50**, 699–712.

Pyrke A (1994). Soil disturbance by native mammals and the germination and establishment of plant species. PhD thesis, School of Geography and Environmental Studies, University of Tasmania, Hobart.

Rayment GE and Higginson FR (1992). *Australian Laboratory Handbook of Soil and Water Methods.* Inkata Press, Melbourne.

Reid N and Landsberg J (2000). Tree decline in agricultural landscapes: what we stand to lose. In *Temperate Eucalypt Woodlands in Australia: Biology, Conservation, Management and Restoration.* (Eds RJ Hobbs and CJ Yates) pp. 127–166. Surrey Beatty and Sons, Chipping Norton.

Reynolds H (1969). Men of substance and deservedly good repute: the Tasmanian gentry 1856–1875. *The Australian Journal of Politics and History* **15** (3), 61–72.

Reynolds H (2004). *Fate of a Free People.* Penguin Books, Camberwell.

Robson L (1983). *A History of Tasmania: Volume I, Van Diemen's Land from the Earliest Times to 1855.* Cambridge University Press, Melbourne.

Robson L (1991). *A History of Tasmania: Volume II, Colony and State from 1856 to the 1980s.* Cambridge University Press, Melbourne.

Saunders DA, Hobbs RJ and Margules CR (1991). Biological consequences of ecosystem fragmentation: a review. *Conservation Biology* **5**, 18–32.

Schlesinger WH, Reynolds JF, Cunningham GL, Huenneke LF, Jarrell WM, Virginia RA and Whitford WG (1990). Biological feedbacks in global desertification. *Science* **247**, 1043–1048.

Scott P (1955). Transhumance in Tasmania. *New Zealand Geographer* **7**, 157–172.

Scott P (1965). Farming. In *Atlas of Tasmania.* (Ed JL Davies) pp. 58–65. Lands and Survey Department, Hobart.

Sharp BR and Whittaker RJ (2003). The irreversible cattle-driven transformation of a seasonally flooded Australian savanna. *Journal of Biogeography* **30**, 783.

Shepherd RR (1973). Land use on the Central Plateau with special reference to the grazing industry. In *The Lake Country of Tasmania: A Symposium Conducted by the Royal Society of Tasmania at Poatina, Tasmania, 11–12 November 1972.* (Ed MR Banks) pp. 161–176, Royal Society of Tasmania, Hobart.

Smith JA (2002). Seed banks, community dynamics and species persistence in five temporary wetlands. PhD thesis, University of Tasmania, Hobart.

Solomon RJ (1972). *Tasmania.* Angus and Robertson, Melbourne.

Strzelecki PE de (1845). *Physical Description of New South Wales and Van Diemen's Land: Accompanied by a Geological Map, Sections and Diagrams, and Figures of the Organic Remains.* Longman, Green, Brown and Longmans, London.

Sutherland WJ (2002). Openness in management. *Nature* **418**, 834–835.

Taylor VC (1985). *Winton Merino Stud 1835–1985.* Neptune Press, Geelong.

Thomas I (1992). The Holocene archaeology and palaeoecology of northeastern Tasmania. PhD thesis, University of Tasmania, Hobart.

Thomas I and Hope G (1994). An example of Holocene vegetation stability from Camerons Lagoon, a near treeline site on the Central Plateau, Tasmania. *Australian Journal of Ecology* **19**, 150–158.

Tiver F and Andrew MH (1997). Relative effects of herbivory by sheep, rabbits, goats and kangaroos on recruitment and regeneration of shrubs and trees in eastern South Australia. *Journal of Applied Ecology* **34**, 903–914.

Tongway D and Hindley N (2000). Assessing and monitoring desertification with soil indicators. In *Rangeland Desertification.* (Eds O Arnalds and S Archer) pp. 89–98. Kluwer, Dortrecht.

Tschamtke T and Greiler HJ (1995). Insect communities, grasses and grasslands. *Annual Review of Entomology* **40**, 559–585.

Valderrabano J and Torrano L (2000). The potential for using goats to control *Genista scorpius* shrubs in European Black pine stands. *Forest Ecology and Management* **126**, 377–383.

Vanclay F, Howden P, Mesiti L and Glyde S (2006). The social and intellectual construction of farming styles: testing Dutch ideas in Australian agriculture. *Sociologia Ruralis* **46**, 61–82.

Vanclay F, Mesiti L and Howden P (1998). Styles of farming and farming subcultures: appropriate concepts for Australian rural sociology? *Rural Society* **8**, 85–107.

Venning J (1988). *Growing Trees for Farms, Parks and Roadsides: A Revegetation Manual for Australia.* Lothian Publishing, Melbourne.

Vercoe J (2003). Voluntary conservation on private land: Tasmania's private reserve program lessons and opportunities. BSc (Hons) thesis, University of Tasmania, Hobart.

Vesk PA and Westoby M (2001). Predicting plant species' responses to grazing. *Journal of Applied Ecology* **38,** 897–909.

Walpole S, Lockwood M and Miles CA (1998). 'Influence of remnant vegetation on property sale price'. Johnstone Centre Report No. 106, Johnstone Centre, Albury.

Wells TCE (1969). Botanical aspects of conservation management of chalk grasslands. *Biological Conservation* **2**, 36–44.

Weltzin JF, Archer S and Heitschmidt RK (1997). Small mammal regulation of vegetation structure in a temperate savanna. *Ecology* **78**, 751–763.

West J (1852) (reprinted 1971). *The History of Tasmania.* (Ed AGL Shaw) Angus and Robertson in association with the Royal Australia Historical Society, Sydney.

Whalley W (2005). Grassland regeneration and reconstruction: the role of grazing animals. *Ecological Management and Restoration* **6**, 3–4.

Wilson D (2004). Vegetation change in the upper Macquarie catchment, between 1968 and 2002: patterns and causes. GradDipSIS (Hons) thesis, School of Geography and Environmental Studies, University of Tasmania.

Woolley A and Kirkpatrick JB (1999). Factors related to condition and rare and threatened species occurrence in lowland, humid basalt remnants in northern Tasmania. *Biological Conservation* **87,** 131–142.

Yates JJ (1973). Introduced pasture species in Tasmania on the Central Plateau. In *The Lake Country of Tasmania: A Symposium Conducted by the Royal Society of Tasmania at Poatina, Tasmania, 11–12 November 1972*. (Ed MR Banks) pp. 153–160. Royal Society of Tasmania, Hobart.

Zacharek AR (1990). Bush invasion by *Cotoneaster* spp. in Hobart. GradDipEnvSt. (Hons) thesis, Department of Geography and Environmental Studies, University of Tasmania, Hobart.

Zacharek AR (1997). Management and restoration of native grassy woodlands in the Midlands of Tasmania. PhD thesis, University of Tasmania.

Zacharek AR, Gilfedder L and Harris S (1997). The flora of Township Lagoon Nature Reserve and its management. *Papers and Proceedings of the Royal Society of Tasmania* **131**, 57–66.

Index